Interacting with Information

Synthesis Lectures on Human-Centered Informatics

Editor

John M. Carroll, *Penn State University*

Human-Centered Informatics (HCI) is the intersection of the cultural, the social, the cognitive, and the aesthetic with computing and information technology. It encompasses a huge range of issues, theories, technologies, designs, tools, environments and human experiences in knowledge work, recreation and leisure activity, teaching and learning, and the potpourri of everyday life. The series will publish state-of-the-art syntheses, case studies, and tutorials in key areas. It will share the focus of leading international conferences in HCI.

Interacting with Information

Ann Blandford and Simon Attfield

www.morganclaypool.com

ISBN: 9781608450268   paperback
ISBN: 9781608450275   ebook

DOI 10.2200/S00227ED1V01Y200911HCI006

A Publication in the Morgan & Claypool Publishers series
SYNTHESIS LECTURES ON HUMAN-CENTERED INFORMATICS

Lecture #6
Series Editor: John M. Carroll, *Penn State University*
Series ISSN
Synthesis Lectures on Human-Centered Informatics
Print 1946-7680   Electronic 1946-7699

Interacting with Information

Ann Blandford and Simon Attfield
University College London

SYNTHESIS LECTURES ON HUMAN-CENTERED INFORMATICS #6

MORGAN & CLAYPOOL PUBLISHERS

ABSTRACT

We live in an "information age," but information is only useful when it is interpreted by people and applied in the context of their goals and activities. The volume of information to which people have access is growing at an incredible rate, vastly outstripping people's ability to assimilate and manage it. In order to design technologies that better support information work, it is necessary to better understand the details of that work. In this lecture, we review the situations (physical, social and temporal) in which people interact with information. We also discuss how people interact with information in terms of an "information journey," in which people, iteratively, do the following: recognise a need for information, find information, interpret and evaluate that information in the context of their goals, and use the interpretation to support their broader activities. People's information needs may be explicit and clearly articulated but, conversely, may be tacit, exploratory and evolving. Widely used tools supporting information access, such as searching on the Web and in digital libraries, support clearly defined information requirements well, but they provide limited support for other information needs. Most other stages of the information journey are poorly supported at present. Novel design solutions are unlikely to be purely digital, but to exploit the rich variety of information resources, digital, physical and social, that are available. Theories of information interaction and sensemaking can highlight new design possibilities that augment human capabilities. We review relevant theories and findings for understanding information behaviours, and we review methods for evaluating information working tools, to both assess existing tools and identify requirements for the future.

KEYWORDS

information interaction, information seeking, information journey, human–centred information retrieval, amplifying human capabilities, human–computer interaction, CSII, CASSM, sensemaking, visual analytics, serendipity

Contents

Acknowledgments

We are grateful to the many colleagues, local and international, who have contributed to this work in various ways, particularly Anne Adams, George Buchanan, Abdigani Diriye, Sarah Faisal, Stephann Makri, Hanna Stelmaszewska, Claire Warwick, colleagues at Lexis Nexis Butterworths, and the Greenstone team at Waikato. Our research in this area has been funded by EPSRC and ESRC.

Ann Blandford and Simon Attfield November 2009

Preface

Human–Centred Informatics (HCI) has grown out of an interest in understanding how people interact with computer systems and how to better design systems to support their users. With a growing emphasis on information work and on the contexts within which particular technologies are situated, it is important to understand people's situated interactions with information, in order to help us understand how to better support them. There are many aspects of the situation, including the physical situation in which interaction takes place, the ecology of resources to which individuals have access, the social structures within which people interact with information, the individual knowledge and skills people bring, and temporal aspects, through which the above can dynamically evolve and change.

Many aspects of information working are "invisible:" people engage in them without recognising their importance or their sophistication. Consequently, little effort has been put into designing systems that support the rich variety of information activities that people engage in. The "query–response" paradigm of information provision is well established, and it has delivered many benefits and newer interaction paradigms such as those supported by Web2.0 technologies facilitate novel forms of information provision and computer-mediated communication. However, these technologies address only a fraction of the information challenges and opportunities that people encounter in their everyday lives.

This lecture makes visible many aspects of information interaction that can often go unnoticed, and it discusses ways of designing and evaluating novel technologies to support those interactions. It draws on work within Information Retrieval, Information Seeking, Human–Centred Informatics, Digital Libraries, Visual Analytics and Sensemaking to present the information journey framework for understanding situated information interactions. It goes on to discuss approaches to the design and evaluation of systems to support information interactions. The target audience is students, researchers and practitioners with an interest in the theory and practice of designing systems to support information interactions, whether coming from information sciences, human–computer interaction or a related discipline.

Ann Blandford and Simon Attfield
November 2009

CHAPTER 1

Introduction: Pervasive Information Interactions

We live in an "information age," but information is only useful when it is interpreted and applied by people in the context of their goals and activities. The volume of information to which people have access is growing at an incredible rate, vastly outstripping people's ability to find, assimilate and manage it. If we are to design technologies that support people's information working well then we need to understand those working practices and the situations in which they occur.

Information seeking is rarely an end in itself, nor is it something that people often plan. In the morning, when people contemplate what they are going to do for the day, "looking for information" is unlikely to feature high on their list. Looking for information is commonly "invisible work" (Daniels, A., 1987), necessary to support the visible but not a focus in its own right. It is part of a broader activity. Within Information Studies, a distinction is made between Work Tasks and Search Tasks, corresponding to the broader activity and the information seeking activity but, as Vakkari, P. (2003) points out, the term "task" is ambiguous, so in this lecture we talk more in terms of activities, goals and processes.

People encounter information in all their daily activities – whether reading the cereal box over breakfast or checking the train times at the station, looking up someone's address or gathering background material for a future presentation. Sometimes, the encounter is deliberate (e.g., picking up a magazine or surfing to a particular web page); other times, it is unsolicited. Sometimes, it involves developing a strategy for finding particular information, whether a simple fact or a more complex set of interrelated concepts. Often, new information will build on that which has been previously acquired to give a richer understanding of a situation.

To illustrate the diverse nature of much information interaction, let us consider a brief example. Planning a holiday typically involves many forms of information interaction, particularly, if the holiday is the first visit to a far-flung corner of the globe. The initial decision to go so far might have been based on various snippets of information encountered effortlessly over many years: a photo in a magazine, a travel article in a newspaper, or some anecdotes from a friend who had been there. There will probably be some explicit searching to find out how much it is likely to cost (by searching for air fares on the Web and looking up the cost of living in a travel guide). There will also be searching and browsing to better understand what is possible and desirable, what distances are involved, the nature of the possible places to visit, what the weather is likely to be, etc. The discovery of some information may lead to the desire to find other information that helps to build up understanding. This process involves finding some particular facts, but also "information grazing," in which new information is acquired and built on without the need to make explicit relevance judgements. This is a form of sensemaking, in which understanding of a situation (in this case, about holiday possibilities) develops iteratively through interacting with information, developing understanding, forming new questions and seeking new information to fill gaps. The sensemaker may oscillate between focused searching and exploratory interactions, since some information needs are well defined while others cannot be clearly articulated (and may not even be recognised). That information is then used in various ways: to buy tickets, plan itineraries, or perhaps to write a blog of the journey as it unfolds.

This brief example illustrates that people interact with information in many ways: sometimes intentionally, sometimes by chance; sometimes to seek particular facts, sometimes to build up an understanding; sometimes with well-defined search criteria, and sometimes without a clear understanding of what information might be available and might be useful. Significantly, the acquisition of information also supports people's goals – whether these are goals that result in some external change (such as making a flight booking) or goals of developing understanding or feeling entertained and informed. Some of these information behaviors are well supported by existing technologies; others are not.

This lecture is concerned with the ways that people interact with information and how those interactions are designed, particularly where they involve digital technologies. We focus on the person at the centre of an "information universe," their activities and needs and, in particular, the roles of interacting with information to support broader activities. We also consider how to design to support information interaction and how to evaluate systems that support information work.

Information interaction takes place within a context that is physical, social and activity-centred, building also on the prior knowledge and experience of the individual or community. Following a review of research related to information interaction, we consider information interaction from different angles (physical, temporal and social). We then present and exemplify the information journey as a way of thinking about information interactions. Moving from theory to practice, we discuss examples of systems that support aspects of information interaction and requirements for future systems; we then present approaches to identifying user-centred system requirements and evaluating current systems against those requirements. Finally, we look to the future of interacting with information.

C H A P T E R 2

Background: Information Interaction at the Crossroads of Research Traditions

As discussed above, information interaction takes places within the context and in the service of some broader activity. That broader activity might be planning a journey or dealing with a health issue (e.g., deciding whether or not to consult a doctor); preparing a legal case or writing an article. The broader activity might be classed as "work" or "leisure;" it might take place in one place (home, street, office) or many. With the rise of connected technologies, it might take place on the move. The activity involves interacting with information in many forms and via many channels (other people, and physical and digital media), mediated by various technologies. Through these many ongoing interactions, the individual experiences an "information journey," in which they recognise a need for information, acquire and interpret the information, and use the interpretation. The ways in which this notion of an "information journey" plays out in practice and how systems can be better designed to support the many information journeys that people go on, are a central focus of this lecture.

This view contrasts with and, yet, complements those which are prevalent within information retrieval, information seeking and personal knowledge management research. To set the context, we briefly review these complementary approaches to studying information interaction. As Ingwersen and Järvelin (2005) note, there has been little communication, historically, between these different communities, and there are great potential benefits to more substantive interactions between them.

In this chapter, we review various approaches that inform our understanding of situated information interaction: information retrieval, information seeking, sensemaking, human–centred informatics, information work, and personal knowledge management. Each of these perspectives contributes to the overall understanding of situated information interaction and how to design to support it.

2.1 INFORMATION RETRIEVAL

Information Retrieval (IR) (van Rijsbergen, C., 2001) focuses on the technologies that support the finding and presentation of information, and places the technology at the centre of the frame. The traditional focus of IR has been on the development of algorithms that improve precision and recall of search results to a submitted query. Precision is a measure of the proportion of search results that are relevant to the query submitted, while recall is a measure of the proportion of the relevant results in the document collection that are returned by the chosen query.

Robertson, S. (2008) presents a history of evaluation in IR. The classical approach to IR evaluation has been the "Cranfield paradigm", dating back to the late 1950s, within which as many variables as possible (including the database of documents over which retrieval is to be performed and the set of queries) are controlled in order to measure and compare algorithm performance. Tague-Sutcliffe, J. (1992) presents a detailed methodology for conducting an evaluation study within this paradigm, highlighting the importance of validity, reliability and efficiency of IR evaluation studies. The more user interaction is taken into account, the more difficult it becomes to achieve the kinds of validity, reliability and efficiency that have been considered essential to the evaluation of IR systems.

Robertson, S. (2008) articulates one of the central challenges in evaluating IR systems: "On the one hand, we can do experiments in a laboratory, characterized by control and artifice. The control enables us to set up formal experimental comparisons and to expect scientifically reliable answers, confirmed by statistical significance tests (Robertson, S., 2008) On the other hand, we can seek external validity and attempt to observe real world events in their natural setting, which involves waiting for them to happen and minimizing any controls and any observer effects – and therefore get potentially rich but messy and noisy results" (p.447). To achieve a balance between control (and reproducibility) and external validity, the focus of research in IR has extended to cover Interactive IR (IIR) systems. In evaluations of IIR systems, user behaviours have often remained controlled. For example, although Borlund, P. (2003) notes that users' information needs are individual and change over time, and that the relevance of search results should be assessed against the need (rather than against the query, as is done in traditional IR evaluation), she presents an IIR evaluation approach that considers three main factors: the components of an experimental setting designed to promote experimental validity (such as simulated work tasks); how to apply simulated work tasks within the experimental setting; and a richer set of performance measures (beyond precision and recall), namely relative relevance and ranked half-life. The details of these measures are not important here beyond noting that, like the traditional measures, they are summative measures of system performance.

Another example of such an approach (of formulating hypotheses and devising controlled experiments that minimise unwanted variability) can be found in the work of White et al. (2006), who studied the effectiveness of an implicit feedback mechanism devised to assess document relevance. In their case, they developed alternative system prototypes, and allocated predetermined tasks to participants, though the measures captured during and after task performance were subjective (e.g., the participants' sense of whether they had completed the task successfully).

Although rarely stated as such, IR systems and their evaluation paradigms make a number of necessary assumptions about user behaviour when looking for information. They assume that users are able to define the information they need in order to address a given problem (even if their needs evolve over time); they assume that users know how to use a query language to formulate a corresponding query to submit to a search engine; and they assume that users are able to recognise the relevance of the results they get with respect to the problem. While these have been useful assumptions for motivating research on increasingly effective search algorithms, they are a poor reflection of the spectrum of information interaction behaviours as sketched above.

2.2 INFORMATION SEEKING

Information Seeking (IS) builds on a library sciences tradition rather than computer science. Information Seeking (IS) is primarily concerned with the seeking of information (Ellis and Haugan, 1997; Marchionini, G., 1995), placing the individual and their finding activities at the centre of the frame. Despite calls to locate information seeking research within rich accounts of context (see, for example, Dervin and Nilan (1986)), there is scope for developing more complete explanatory accounts in terms of the information seeking context, or situation, and of understanding the reciprocal interplay between IS and contextual factors such as the evolution of wider activities and understanding.

Various IS models have been proposed, some focusing on the temporal aspects, others on the behaviors that individuals exhibit while looking for information.

One theme that has emerged through many studies is the iterative process through which an individual moves while seeking information. For example, Marchionini, G. (1995) describes an Information Seeking Process that has eight stages:

1. Recognition and acceptance of an information requirement.

2. Definition of the information problem.

3. Selection of an appropriate source which might address the problem.

4. Formulation of a query.

5. Execution of the query.

6. Examination of query results.

7. Extraction of information from result documents.

8. Reflection on the process.

Iteration is at the heart of this process; for example, examination of query results may lead the user to reformulate their query. Marchionini also highlights the opportunistic aspect of searching, as the user identifies new opportunities in the information being retrieved.

This theme of identifying new opportunities is also central to the "berrypicking" model of Bates, M. (1989). In this view, the information seeker starts with a need (which may be poorly formulated) and as they interact with information sources they pick up both resources ("berries") and also new ideas of ways to further the search, and so the search evolves over time as information triggers shifts in thinking and new lines of enquiry.

Drawing on a similar ecological analogy, Pirolli and Card (1995) describe information seeking in terms of "foraging." The foraging analogy provides an account of how people, while browsing information resources, choose to continue browsing in the same "region" or choose to identify a new region in which to look for information. They use this understanding to propose a novel approach to designing the interaction between user and information resources, based on "information scent" (Chi et al., 2001), which they describe as the quality of the information pointers that indicate the likely substance of an article (e.g., a web page) and allow the user to assess its interest to them.

With an interest in describing and ultimately designing for naturally occurring behavior, Ellis and colleagues (e.g., Ellis, D. (1989); Ellis et al. (1993)) identify eight primary information behaviors:

1. Starting: identifying sources of interest.

2. Chaining: following leads from an initial source.

3. Browsing: scanning documents or sources for interesting information.

4. Differentiating: assessing and organising sources.

5. Monitoring: keeping up-to-date on an area of interest by tracking new developments in known sources such as journals.

6. Extracting: identifying (and using) material of interest in sources.

7. Verifying: checking the accuracy and reliability of information.

8. Ending: concluding activities.

Apart from the first and last behaviors, there is no implied ordering of activities in Ellis' model. In contrast, the Information Search Process (ISP) model of Kuhlthau, C. (1991), which is based around an information task such as an essay assignment, identifies six stages of the information search process through which an information seeker moves on the path from initial uncertainty, through exploration, to understanding. According to Kuhlthau, C. (1991), early on in a new search, users may not even know what they are looking for: problem formulation emerges through the dialogue between a partially specified topic (e.g., an essay title) and a set of information sources, and the user may have difficulty assessing the relevance of documents to their information task. A browse-based interaction can be more effective than search at this stage, depending on the thematic organisation of the information sources. As the task becomes better understood, the bounds of the information problem become clearer, and the issue becomes more one of understanding how to formulate a query for this particular search interface. Then, when browse or search results are returned, the user has to be able to assess the relevance of each document rapidly and reliably.

The idea that information seekers are frequently uncertain about the information they want has been a persistent theme within information seeking research. Taylor, R. (1968) defines different levels of information need within a process of moving from an actual but perhaps unrecognised need to an expression of a need which could be presented to an information system. Belkin, Oddy and Brooks' ASK hypothesis (Belkin et al., 1982a,b) echoes this idea by stating that "an information need arises from a recognized anomaly in the user's state of knowledge concerning some topic or situation and that, in general, the user is unable to specify precisely what is needed to resolve that anomaly" (Belkin et al., 1982a, p. 62).

Historically, research on information seeking has been based around the physical library, more recently moving to consider interactions with the World Wide Web and digital libraries. The focus remains largely, however, on the information seeking activity. Although context has become more significant, there remains a need to develop a greater understanding of the coupling between information seeking and the broader situation in which it occurs.

2.3   SENSEMAKING

In terms of making this link between evolving IS behavior and the evolving situation in which it arises, a promising perspective is one that locates information interaction within sensemaking. Sensemaking locates the focus of interest on *why* people are seeking information and what is being done with that information. It has been described as "the reciprocal interaction of information seeking, meaning ascription and action" (Thomas et al., 1993, p. 240) and as "the deliberate effort to understand events" (Klein et al., 2007, p. 114). It occurs when people face new problems in unfamiliar situations and their current knowledge is insufficient (Zhang et al., 2008).

The study of sensemaking has been conducted, apparently independently, in Naturalistic Decision Making (e.g., Klein et al. (2006)), Organisational Studies (e.g., Weick, K. (1995)), Information Science (e.g., Dervin, B. (1983); Savolainen, R. (2006)) and Human–Computer Interaction (e.g., Russell et al. (1993); Pirolli and Card (1995)). Whilst these various approaches emerge from the study of human activity in diverse natural settings (such as military command and control, healthcare, commercial organisational life and training course design), they share a common interest in mapping out the processes through which people construct meaning from the information they experience. A common characteristic which has been identified in various studies of sensemaking and which could be said to be a signature phenomenon is an interplay that occurs between top-down and bottom-up processing. Specifically, sensemaking operates as a bi-directional process under the influence of data on the one hand and the generation of representations that account for data on the other.

Klein et al. (2006, 2007) present this interplay in terms of their data-frame theory of sensemaking. This theory presents sensemaking as a continual process of framing and re-framing in the light of data. As someone encounters a new situation, a few key elements invoke a plausible internal representation (a "frame") as an interpretation of that situation. Active exploration guided by the frame then elaborates it with more supporting evidence or challenges it by revealing inconsistent data. A frame offers economy in terms of the data required for understanding, but it also sets up expectations of further data that might be available. Hence, a frame can direct information seeking and, in doing so, reveal further data that changes the frame. Klein et al. (2007) argue that an activated frame acts as an information filter, not only determining what information is subsequently sought but also what aspects of a situation will be noticed.

The symbiotic interaction between data and frame also features prominently in the sensemaking account of Weick, K. (1995). He argues that when people make sense of information, they do so by placing it into a framework which allows them to categorise, fill in missing data, assign likelihoods to data, and filter and hide data.

Within Human Computer Interaction and Information Science, research has focused on technologically mediated sensemaking. The relevance of sensemaking arises from the fact that users interact with an information system in order to develop some 'picture' or 'model' of a domain (Dervin, B., 1983; Spence, R., 1999). Technologically, mediated sensemaking often involves searching for and integrating large amounts of information into a coherent understanding. Whereas, for both Klein and Weick, the representations considered are internal and cognitive, within Human Computer Interaction, there has been a particular interest in the role and design of technologically supported, user-generated externalisations of domain representations (e.g., Russell et al. (1993); Pirolli and Card (2005)). Despite this difference, the same bi-directional process between data and representation is evident. For example, Pirolli and Card (2005) report preliminary findings from a study of intelligence analysts, which exemplifies the interplay between top-down and bottom-up processing in sensemaking. Their model shows transformations that the analyst performs in converting multiple data sources into novel information. It consists of two major activity loops: a foraging loop and a sensemaking loop. Foraging involves seeking information, filtering it, and reading and extracting information, possibly into some schema. The sensemaking loop involves the iterative development of a "mental model" or "conceptualisation" from the schema that best fits the evidence. The model is not committed to a single direction of processing: there is an opportunistic interplay between both kinds of process. From bottom to top, the analyst searches or monitors incoming information and sets aside relevant information as it is encountered, then nuggets are extracted and re-represented schematically, and a theory develops and is ultimately presented to some audience. In the opposite direction, new theories suggest hypotheses to be considered and the schemas are re-considered in this light, collected evidence is re-examined, new information is extracted from stored information, and new raw data is sought.

Reflective of the different disciplinary contexts within which sensemaking research has been conducted, there have been different motivations for the research, including seeking a better understanding of cognition and designing improved systems to support sensemaking. Much of the latter has been in a Human–Centred Informatics tradition. For example, Stasko et al. (2008) build on sensemaking theory, including the work of Pirolli and Card (2005), to develop a system that supports intelligence analysts in making sense of large bodies of information. We return to this theme of designing novel systems to support information work in Chapter 5.

2.4   A HUMAN–CENTRED INFORMATICS VIEW

The design, development and testing of sensemaking support tools is one example of Human–Centred Informatics research related to information interaction. HCI is concerned with both the design and the evaluation of interactive systems.

On the design side, there have been a range of novel interaction designs, including visualisations of information structures (e.g., Shen et al. (2006)) and page-turning systems that mimic the behaviour of physical books (e.g., Chu et al. (2004)), and systems to support specific user needs (e.g., Morris et al. (2004)). The space of possibilities is enormous; in many cases it is unclear to what extent systems have been developed in response to established user requirements and to what extent they are the fruits of investigations into what is technically possible. In chapter 5, we discuss various designs to support the information journey that are explicitly based on studies of people's information behaviours and requirements.

Much of the work in HCI for information interaction has focused on evaluation – e.g., of particular digital libraries and similar information repositories. A variety of methods have been employed, addressing a range of evaluation questions.

Expert (or analytical) evaluation involves experts (typically HCI experts rather than domain experts) assessing a system without the involvement of the intended users of the system. As an example of expert evaluation, Hartson et al. (2004) studied the Networked Computer Science Technical Reference Library (http://www.ncstrl.org/). In their study, evaluators employed a co-discovery method, described by Hartson *et al.* as an approach in which two or more usability experts work together to perform a usability inspection of the system. The resulting verbal protocol forms the basis of the usability evaluation, which focused primarily on usability problems with the system interface. Blandford et al. (2004) applied four different analytical techniques, Heuristic Evaluation (Nielsen, J., 1994), Cognitive Walkthrough (Wharton et al., 1994), CASSM (Blandford et al., 2008a) and Claims Analysis (Carroll and Rosson, 1992) to various digital libraries (DLs). They found that Heuristic Evaluation and Cognitive Walkthrough only address superficial aspects of interface design (but are good for that), whereas Claims Analysis and CASSM helped to identify deeper conceptual difficulties (but demanded greater skill of the analyst), because they force a deeper engagement with the domain of activity (i.e., how users interact with the information in the digital library, and understand the system with which they are working). This is a topic to which we return in chapter 6.

Another approach to understanding the use of web-based information-based systems has been the use of transaction logs. For example, Mahoui and Cunningham (2000) compared the transaction logs for two collections of computer science technical reports to understand the differences in searching behaviour and relate that to the different designs of the two systems. Mahoui and Cunningham (2000) argue that the value of transaction logs lies in the availability of data about a large number of transactions and users, making it possible to develop an understanding of behaviour with a particular system (though not of particular users).

Some studies have used "surrogate users" – subject experts who can better assess features of a DL than the target user population, but who are not usability experts. One example is the work of McCown et al. (2005), who recruited eleven teachers to participate in a study comparing the effectiveness of the National Science Digital Library (http://nsdl.org/) and Google in terms of the quality of results returned for curriculum-related search expressions. In this case, the evaluation did not assess the quality of the interaction or system design, but considered the quality of the results returned in relation to the relevant school curriculum.

Most studies involve current or potential system users. While some user studies have been quantitative, in a similar tradition to the IR studies described above, more have been qualitative, based on think-aloud, interview or observation. Some studies are clearly in an HCI tradition, evaluating the interface, whereas others, in seeking to understand the broader context within which users' information work takes place, span traditions of HCI, Information Seeking and Information Working. For example, Blandford et al. (2001) conducted a study of how Computer Science researchers work with multiple DLs using a think-aloud protocol, focusing on the interactive behaviours regardless of the broader context, while Kuhlthau and Tama (2001) used semi-structured interviews to study the information practices and needs of lawyers. Blandford *et al.* used an observational technique because what mattered was what people actually do rather than what they think they do; in contrast, Kuhlthau and Tama used interviews because their focus was on lawyers' perceptions rather than the details of their information behaviour. In the latter case, general features of existing systems were considered, rather than the details of particular system designs.

2.5   INFORMATION WORK

The work of Kuhlthau and Tama (2001) is an example of a study of information practices and needs within a particular professional setting. One aim within both HCI and Information Seeking research has been to understand interactions as these arise from the working context within which they take place. Examples include Ellis and colleagues' studies of the information behaviors of researchers in the physical and social sciences (Ellis et al., 1993), of research scientists and engineers (Ellis and Haugan, 1997) and of English Literature scholars (Ellis and Oldman, 2005). Makri et al. (2008) validated and extended Ellis' work by investigating the information behaviors of lawyers.

Although they studied a range of professional groups, Ellis and colleagues were primarily interested in identifying information seeking behaviors (rather than information interactions in general) and did not relate these back to the evolving task. Others have expanded this focus to include the ways in which information is subsequently managed, organised and used over the course of an evolving work-activity.

For example, Attfield and Dowell (2003) studied the information practices of journalists, considering not just how they found information, but also how they subsequently interacted with it in the construction of a news report. They highlighted the constraints under which journalists worked (such as the need to ensure accuracy of their reports, and to develop an angle for an article), which gave shape to the ways in which they found, evaluated and managed information. They also described how information seeking and information gathering allowed the journalists to develop a "physical and cognitive resource space" (p. 201) that helped them address the task. This took the form of task-oriented information collections in electronic and paper-based form supplemented by their personal knowledge. Significantly, the nature of the task was prone to change mid-assignment (constraint changes) and this had implications for how journalists interacted with information tools and collections.

Retaining an interest in the way that users maintain and develop task resources, but with more focused attention on the activities of a single journalist writing a single article, Attfield et al. (2008b) noted a pattern of behavior in which the journalist repeatedly immersed herself in information in order to generate and document ideas, followed by a period of consolidation in which the resulting artefacts were structured to provide a resource that would optimally support the next stage of immersion, until the final article was written and submitted for publication.

Palmer, C. (1999) studied the work practices of interdisciplinary scientists, focusing particularly on their strategies for working across disciplinary boundaries. She highlights some information practices that are common to all scientists, and others that are needed to address the demands of interdisciplinary working; for example, she highlights the need to rapidly assimilate bodies of research outside the core area of expertise, and to make sense of information from another discipline that might be written using different language, or with different basic assumptions.

Taking a broader perspective, Palmer et al. (2009) considered the activities of scholars, and identified five key activities: searching; collecting; reading; writing and collaborating. These broad information behaviors all involve four cross-cutting primitives: monitoring; note-taking; translating; and data practices. They propose that it is a role of the library (through both physical and digital infrastructure) to support this broad range of information behaviors, rather than the more narrowly conceived role of simply providing information.

2.6   PERSONAL KNOWLEDGE MANAGEMENT

Other fields such as "personal information management" and "knowledge management" also touch on how individuals interact with information. Work in knowledge management typically takes a broader organisational perspective. McAdam and McCredy (2000) summarise definitions of "knowledge management;" taking a holistic view, it might refer to the creation, interpretation, dissemination, retention, refinement and use of knowledge, whereas a more mechanistic view might focus on the management of human-centred assets. The broader of these views is relevant to information interaction as it highlights the outcomes of working with information. Wilson, T. (2002) argues that it is "nonsense" to talk about "knowledge management," since knowledge resides in the head, and, therefore, that which can be managed is "information."

Taking a more "bottom up" approach to understanding personal information management, a number of researchers have focused specifically on the way individual knowledge workers gather and use information. For example, Sellen et al. (2002) reported a diary study of Web activity of knowledge workers over a two-day period. They discuss 'finding' and 'gathering.' Finding corresponds to locating specific facts which might be maintained for reference on a temporary basis. 'Gathering' involves locating and storing information to address questions which were difficult to specify. Once gathered, information is stored in a number of ways, although printing is preferred since this allows documents to be close to hand and provides task context. Implicit within an information storage strategy is the need for relocation of that information. Where a lot of information is involved, users may promote easier relocation by creating an organisational schema (Jones et al., 2002). Hyams and Sellen (2003) found that personal information collections tended to be organised by project or by topic, and gathered information was rarely used outside the boundaries of a project. Such research shows that the creation of personal information collections is an important component of knowledge work. Knowledge workers develop tactics for ensuring that, as they encounter useful information, they can create ways of re-encountering it later, in response to particular task demands. Motivated by such observations, Dumais et al. (2003) built on an understanding of how people retrieve previously seen information in the development of "Stuff I've Seen," an information management system that helps people exploit their memory of what context information was previously found in as an aid to refinding.

2.7   SITUATED INFORMATION INTERACTION

In this chapter, we have reviewed many of the approaches that contribute to an understanding of information interaction. Each has a different focus: on the underlying algorithms for information retrieval, the human behaviors of information seeking, the mental processes of making sense of information, the qualities of the interaction between a person and a system, or the broader activity within which information is used. In subsequent chapters, we focus more explicitly on people's interactions with information: understanding the situations within which information interactions take place (physical, social and temporal) and the more detailed process of engaging with and using information in the construction of new knowledge and action. We present an "Information Journey" view that focuses on the unfolding interactions between people, information, and the systems that make that information available and support its manipulation. We then discuss approaches to designing to support the information journey, and we discuss methods for evaluating the systems that result.

CHAPTER 3

The Situations: Physical, Social and Temporal

Information interaction always takes place within some setting. While that setting is inherently whole (e.g., the social structures influence and are influenced by physical structures and artefacts, and social structures co-evolve with practices), we consider it here from three perspectives: the physical setting, the social situation and the temporal evolution of practices.

3.1 THE PHYSICAL AND THE DIGITAL

The interrelationships and interplay between the physical and the digital are most evident in the places where people find and use information, and the nature of the information resources that they use. In terms of where people find information, we focus on the "library" and its changing role; in terms of resources, we focus on information artefacts such as books and papers.

3.1.1 THE PHYSICAL AND THE DIGITAL LIBRARY

The advent of the World Wide Web and, more recently, mobile computing have created a revolution in the ways that people interact with information. The ability to access rich information resources from almost any place, at almost any time, have transformed the ways that people work, and the temporal coupling between information seeking and information use.

Historically, it was generally necessary either to have selected an information resource to take to the workplace (either buying it or borrowing it from a library), or to go to the library to find information resources. In this situation, it was advisable to go to the library with a fairly well formed view of what one was looking for and to be able to form accurate relevance judgements of material before copying or borrowing it. Arguably, much of the work on relevance assessment is predicated on the view that there is a significant cost (in time or resources) to acquiring information, and that it is, therefore, important to assess the relevance of information resources before investing in acquiring them.

With the new availability of information in the work setting, there is a much closer coupling between information finding and information use. Consequently, it is no longer necessary to plan for finding and to dedicate time to this activity, including moving to a different physical space to do so. This makes possible different information behaviors: it is now possible to "graze" for information, interleaving finding and using seamlessly (Twidale et al., 2008). This has consequences for both the role of the library and the role of the librarian; we consider the former here and the latter in a later section.

There are many definitions of what a digital library is (Borgman, C., 2003), but the key features these definitions share is that documents are selected for inclusion in the library (hence there is some form of quality control or other selectivity criterion), that there is an organising principle for the presentation of documents, and that the library takes responsibility for preservation of content over time. These are features that digital libraries share with their physical counterparts.

Duncker, E. (2002), however, uses a study of Maori perceptions of libraries – both physical and digital – to argue that the analogy between physical and digital libraries is of limited value. Although digital and physical libraries have features in common, the use of the library metaphor – for example, in classifications of documents and in the browsing structures implemented – and the lack of support for use were found to make DLs essentially unusable by Maoris. She points out that while physical libraries are well embedded in many cultures, there are others – such as the Maori culture – that have an oral tradition, using trusted individuals as "living repositories" (p. 224) of tribal knowledge. Since libraries are repositories of knowledge, they are typically viewed as sacred – a view that is in tension with the fact that library-based knowledge is publicly accessible. In addition, Western-style subject headings and classification systems are meaningless to such cultures. The cultural practices around the use of a physical library are many and subtle.

Harrison and Dourish (1996) discuss the important distinction between *space*, the physical structure inhabited, and *place*, the "sets of mutually-held, and mutually available, cultural understandings about behavior and action." The physical library as a place affords particular kinds of behaviors and interactions.

Stelmaszewska and Blandford (2004) studied the behavior of a group of computer scientists whilst using a physical library. They focused on how they interacted with artefacts, evaluated them and interacted with librarians. Their study highlighted the fact that the library metaphor is limited in terms of how experiences can be transferred from the physical to the digital environment. Users' experiences of using paper, books, shelves and other tangible media are situated and tactile in a way that contrasts with that of using computer keyboards and screens. The digital library is not a simple replacement for the physical one, but affords different experiences and supports different interactions with research materials.

In a study of humanists' use of the library, Rimmer et al. (2008) found that people placed great store on the importance of, and authenticity of, the library. The traditional physical library has been described as "the humanists' laboratory" (Burchard, J., 1965): the library materials of the humanities scholar are often their research objects, and their location has a significance that goes beyond simple geography. One participant in the study of Rimmer et al. (2008) talked of the "real experience" of being in the library: of the users being able to "immerse themselves in a particular society," another of feeling "like a historian" when working with old texts in a library. Others talked about the ambience of particular libraries (e.g., one was described metaphorically as a "slum" and an "airport lounge," another as "Stalinist," a third as "a lovely place to work"). The physical sense of community was also important: one participant talked of the "other dusty people" working in the same space and the sense of community that engendered. Another talked about the authenticity of the library, the sense of connection to events that had taken place there in the past, and "tangible immediacy with your subject. Knowing that Carlyle used this library, knowing that Catherine Macaulay used the British Library, knowing that Marx wrote Das Kapital in that room." Conversely, many participants highlighted the advantages of remote access, which creates opportunities to do research without the high costs, both financial and in time, of having to travel to a distant library. They often delighted in the freedom to be able to conduct research from home or their office, without the constraints of being in a particular place at a particular time and without having to deal with so many mundane practicalities such as malfunctioning photocopiers. They also noted how daunting it could be to learn to use a new library, using words such as "scary" to describe the experience. Physical constraints of space could add to the challenge of information access; for example, one participant described a library where "they organise things by size of books. So the folio edition and the Octavo editions are all on particular shelves," while another described the frustrating experience of "where the book is mis-shelved or it is in the wrong place, or it's on the trolley, and how do you know it's on the trolley?"

The physical library may have a particular significance for humanists, but for all library users, it signifies a place to study, and one where there is a possibility of encountering a community of people with similar interests. A library is a place which can be as evocative and culturally nuanced as it can be informative. And it is a space which creates its own constraints and possibilities for information access. Conversely, the digital library offers improved access to a broader range of information, and given that it is *there*, it can offer a much closer coupling between information finding and use, so that information work can be more tightly embedded within the flow of a broader activity.

3.1.2   PHYSICAL AND DIGITAL INFORMATION ARTEFACTS

It might appear that the growing availability of digital information will make physical information artefacts, such as books and journals, obsolete. However, there is plenty of evidence that this is not the case. Based on an ethnographic study of paper usage, Sellen and Harper (2002) highlight the affordances of paper, i.e., "what people can do with paper" (p. 17). These include the following:

- Authoring: writing and annotating are central to composition, as is reference to paper documents while authoring online.

- Reviewing: people were observed to read colleagues' work reflectively on paper, and annotate and comment on it as they did so.

- Planning and reflecting: people were observed to use pen and paper as the primary means of organizing their work.

- Supporting collaborative activities: a paper copy of reference material provides a joint focus for activity.

- Supporting organizational communication: in particular, important documents are handed over to colleagues, signifying receipt as well as transmission in a way that electronic submission does not support.

They also highlight limitations on the use of paper, including the fact that they cannot be remotely accessed, that they take up physical space, that they are difficult to revise and replicate, and that the information displayed is static.

O'Hara et al. (2002) report on the physicality of paper use in a study of participants interacting with multiple source documents during everyday writing tasks. They focused on the movement of attention between documents and composition, how spatial layout is managed to support cognition, and the role of annotation and mark-up. They observed periods of frequent attentional shifts between source documents and composition during writing. During these periods, participants might use a finger or annotations as spatial markers.

O'Hara also observed participants spreading documents out on their desks to enable visual availability without sacrificing visibility of the composition. This layout could change many times as different source documents became the focus of attention. On occasion, participants used paper and electronic versions of the same document, concurrently. For example, they might use a word-processor's search facility to locate specific information, but they might also browse a paper version where they had more 'implicit awareness' of something useful.

Appropriately designed digital presentation media can support key interaction possibilities. For example, Rodden and Fu (2007) report on how people use the mouse to keep track of place in a document, and large display surfaces may support some of the document layout behaviours observed.

The focus of the studies of Sellen and Harper (2002) and O'Hara et al. (2002) is on the objective properties of paper and digital media and on how they are used, singly or in combination. They paid less attention to the different kinds of information that they might contain.

Information may be lost through the digitisation process (e.g., omitting to digitise annotations on the reverse of photographs) or through the absence of modalities (e.g., the feel or smell of the original) in a digitised copy. Davis-Perkins et al. (2005) note that, while the digitisation of photograph collections improves their accessibility, and makes it possible to "improve citizens' sense of self and their society's historical context" (p. 278), the digitisation process itself loses information "through manipulation and accidental loss" (p. 282), which makes the resources of less value to specialists. As well as loss of information, Rimmer et al. (2008), in their studies of Humanists, found that the reliability of the information in digital surrogates (accuracy, quality and credibility) was a concern to several of the participants. In addition, during optical character recognition, the words that are most likely to be incorrectly detected are low-frequency words. Paradoxically, these are the more useful words for information retrieval given a tendency to offer more discriminating characterisations of document context.

Properties such as feel or smell may also contain information not available in a digital document. For example, Rimmer et al. (2008) found that some participants reported on the importance of the materiality of documents for them – e.g., inferring important information from the paper or binding of a book – whereas others were more concerned with content independent of the original context (as long as they could be confident that material had been reliably reproduced).

Just as physical spaces are associated with particular experiences (see above), so physical arte-facts can evoke experiences. There is a particular "magic" in handling very old documents: knowing the provenance of the artefact, having confidence in its authenticity, and having a sense of privileged access are essential elements of that experience.

The experience of interacting with information in physical and digital forms has material and hedonic values that often augment the informational value of the document content. Different kinds of documents support different kinds of interaction that, in turn, support different kinds of information work. Duguid, P. (1996) uses the example of the pencil that still has an important role to argue that technologies that perform the "same" function do not necessarily supersede each other, but may instead be complementary, depending on the task in hand.

In this section, we have noted that people often use physical and digital information resources simultaneously, in complementary ways, and that physical attributes of objects and places also impart information that can be interpreted in the situation of the individual. In practice, in many situations, people also draw on other people as information resources, and develop their understanding of a situation through such interactions, effectively working in a rich ecology of information resources embodied in paper and digital information artefacts, other objects and other people. In the next section, we focus on social aspects of information interaction.

3.2   SOCIALLY SITUATED INFORMATION INTERACTION

The majority of information interactions take place within a social context – whether that be groups of information professionals, peers with similar or different roles or subject matter experts. In this section, we consider some common social situations for information interaction and the roles and relationships of people around information. We start with probably the most widely discussed role related to information work, that of the information intermediary. We then consider two other situations: the evolving "community of practice" around information work and relationships between domain experts and consumers.

3.2.1   INFORMATION INTERMEDIARIES

Historically, librarians have had a formalised role as intermediaries between information consumers and information sources. In both work and leisure settings, other people may also frequently serve such intermediation roles – a topic to which we return in the next section. With the rise of digital information resources, it has been necessary to reconsider future roles of information intermediaries.

Information intermediaries have always had many roles, even when based almost exclusively in physical libraries. Some of these roles (e.g., organising and classifying information) are outside the scope of this lecture. However, in traditional libraries, one important role, particularly for archivists, is as custodian of documents, a role which can effectively turn into one of gatekeeper for the information held in those documents. Rimmer et al. (2008) note that the attitude of an individual archivist can influence the research that is conducted, as ease of access to documents is essential to the conduct of much of the work of humanities scholars. Various authors (e.g., Nardi and O'Day (1996)) argue that direct access to documents, without gatekeeping, is generally beneficial and a potential advantage of digital information resources. However, others (e.g., Makri et al. (2007)) note that people's understanding of access rights to documents in digital libraries and a fear of being charged for access are inhibitors that can discourage people from accessing material to which they do in fact have access rights.

Another important role, with wide impact across disciplines including engineering (e.g., Fields et al. (2004)), law (e.g., Attfield et al. (2008c)), and health (e.g., Adams et al. (2005b)), is that of gathering information for others to use. For many professional information workers, information gathering is peripheral to their main activities, and it can be fruitfully supported by others, such as clinical or legal librarians. But understanding the information needs of others can be a complex matter. Taylor, R. (1968) identified five filters librarians use to understand information needs: subject matter, motivation, personal characteristics, relationship of the enquiry to the file organisation, and what the client anticipates in the form of an answer. Eliciting this information in the reference interview is a significant challenge for the reference librarian. Indeed, the challenge of helping people to articulate and then address their information needs has been described as being akin to psychotherapy (Theng, Y., 2002).

In many situations, the physical reference desk is now being replaced by virtual (or digital) reference services, and the challenge of helping people to articulate and satisfy their information needs is being addressed in new ways. For example, digital (or virtual) reference services use software and the internet to facilitate human intermediation at a distance (Lankes, R., 2004). Question negotiation in this setting can pose particular challenges, as well as new opportunities. For example, Attfield et al. (2008c) highlight the challenges that specialist law librarians face in understanding their clients' needs. In particular, enquirers do not volunteer important information to the service, and asynchronous communication media and some social obstacles present barriers to prompting. They focus on three kinds of information that need to be negotiated between information user and intermediary to support decision making: the priority of the enquiry (needed for time management), what information resources to access (and which may have already been investigated), and suggested search terms (enquirers in this situation were lawyers and therefore domain experts).

A further role for information intermediaries is that of assimilating, merging and re-creating information to suit users' needs (Gristock and Mansell, 1998). One important context in which this occurs is in maintaining current awareness. Attfield and Blandford highlight the roles of knowledge management workers in filtering, selecting, aggregating, and disseminating information to lawyers. They described a complex sociotechnical distribution network which was able to "self-tune" according to local changes in need. Within this network, knowledge management staff acted as a layer of "intelligent filters" supplying information according to their understanding of local information needs—understanding which they acquired through a combination of formal and informal social interactions with lawyers, the majority of which could be arranged or simply arose through being a part of the situation in which the work was done.

Vishik and Whinston (1999) highlight the importance of a trust relationship existing between intermediaries and end users; in this case, their discussion is in the context of reliability of resources. Information intermediaries often have a responsibility for ensuring the quality of information that is available. Information users have variable skills in assessing the quality of information they find; for example, Sillince et al. (2004) found that lay people have difficulty in assessing the quality of health information that is available on the Web. Professionals have more sophisticated strategies for assessing the quality of information, e.g., by cross-reference to information they already know or by carefully selecting reliable sources.

Finally, information intermediaries in many organisations have responsibility for training others to access and use information. For example, Wu and Liu (2003) describe the role of "database instructor," in which the librarian trains users to make use of library databases, independent of the users' personal information needs.

In summary, information intermediaries have a variety of roles in ensuring that users have access to high quality, usable information. While the ways these roles are enacted are changing with the growing use of digital information resources, e.g., with librarians moving out of the library and into the workspace, many of the core responsibilities remain largely unchanged. The need to know the information users and to respond to their needs remains central.

3.2.2   COMMUNITIES OF PRACTICE

As well as sometimes interacting with professional intermediaries, people often interact with other people to find, make sense of and work with information. Sometimes, the relationships are formalised, with clear hand-overs; for example, journalists receive their assignments from an editor and deliver their articles back to the editor (Attfield and Dowell, 2003). In other cases, the relationships are informal and situated; for example, clinicians in a study by Adams et al. (2005b) often reported discussing documents with peers to check their understanding of what was written and what it might mean in a particular clinical context.

These informal information behaviors typically evolve within a "community of practice" (Wenger, E., 1999). According to this view, individuals participate in communities, learning from and contributing to those communities, evolving effective practices in response to new opportunities and needs. Newcomers to a community learn how to participate and may, over time, bring their own prior experience to contribute to new developments in the practices of the community. Adaptations respond to both new possibilities and constraints imposed—e.g., by requirements to be accountable for the process by which information was identified.

An example of how newcomers learn to become part of a community was provided by a participant in the study by Rimmer et al. (2008), who described their early experience of being a PhD student and acquiring new research strategies as being like "osmosis," through observing the supervisor and other academics interacting and finding and working with information.

Adams et al. (2005b) relate the findings of several studies to the theory of communities of practice. They studied four different situations in which information was made available to knowledge workers (clinicians and academics) via digital and traditional libraries. In two of the situations, knowledge workers had access to documents both in a traditional library (staffed by professional librarians) and through digital libraries that were accessible via existing computer systems in people's offices; in one situation, there had been a policy of making information more easily accessible by placing computer systems – and hence digital libraries – in shared spaces (in this case, hospital wards); in the fourth situation, information intermediaries were employed to work with staff and library resources, participating in team meetings and making themselves available to respond to information queries at publicised times. People's perceptions of the technology, and of the usability and utility of information resources, was studied in all four settings.

It was found that participants in the traditional setting (with a physical library and computers in people's offices) perceived the library resources as being irrelevant for their needs; the access to and use of information was only loosely integrated with their community practices (e.g., in taking printouts of papers to meetings for discussion).

Making technology available within shared physical space, but without providing localised support in its use, was widely perceived as a threat to current organizational structures: senior staff were reluctant to use it because they did not feel that they had the necessary skills, while more junior staff were discouraged, or even blocked, from using the systems. Various reasons were presented for this, such as information access not being as important as patient care and needing to password protect computers to prevent unauthorised access (Adams and Blandford, 2002). The new systems disrupted established practices, which generally involved the dissemination of information from the top of the organisation, without facilitating the evolution of new practices.

In contrast, the fourth setting, in which technology could be adapted to and change practices according to individual and group needs, supported by an information intermediary, was seen as empowering to both the community and the individual. In this fourth setting, people reported having the confidence to acquire new information skills, supported by the intermediary, and to develop new practices that enabled them to better keep up with developments within their professional disciplines. With the support of an intermediary, the community evolved new information practices that exploited the new information resources that were made available to them.

One of the real challenges with making information available digitally is that people can, in principle, access it from anywhere. However, many digital libraries are difficult to use (Hartson et al., 2004; Adams et al., 2005b), and without support from other people, whether information professionals or peers, people struggle to use such resources.

Looking to the future, there is a need to better understand the roles of communities of practice and how communities can be empowered to develop new information practices that exploit the possibilities offered by novel information tools.

3.2.3   EXPERTS AND CONSUMERS

As well as needing to develop skills in finding and working with information, people often need to draw on the expertise of domain professionals (doctors, lawyers, etc.). People are evolving new ways of interacting with such professionals as they gain access to more online information resources. There are widespread reports of cyberchondria (e.g., White and Horvitz (2009)), in which patients confront their doctors with internet printouts accounting for their symptoms, and of the "expert patient," who accesses online communities and information to develop a richer understanding of their condition (Fox et al., 2005). Those in favour of disintermediation anticipate the day when lay people can access the information they need without professional help; for example, Susskind, R. (2008) argues that "the market is increasingly unlikely to tolerate expensive lawyers for tasks (guiding, advising, drafting, researching, problem-solving and more) that can equally or better be discharged, directly or indirectly, by smart systems and processes" (p. 2). However, there is growing evidence that access to information is not sufficient to address lay information users' needs.

To illustrate this point: Attfield et al. (2006) report on a study of patients' use of health information. Participants expressed varying levels of confidence in their healthcare professionals' judgements. Some patients expressed concern that their health practitioners might not take account of holistic healthcare factors such as considering potential interactions of treatments with other conditions, or that resource limitations might constrain the extent of tests and treatments offered. Also, patients expressed concern that the practitioner's knowledge might be incomplete or out-dated. Such concerns led patients to seek information both before and after a clinical consultation.

Participants described seeking information prior to a consultation for three main reasons: assessing the need for consultation, deciding who to see, and preparing for the consultation. In preparing for a clinical consultation, participants reported two main aims. The first was to enable them to become a partner in their healthcare by contributing more usefully to the consultation (and thereby reducing demands on Health Service resources). The second was to enable them to be more questioning in the consultation. In both cases, understanding their condition, treatment options, and how these might relate to their own specific circumstances were important.

Following a consultation in which a diagnosis was given, some participants reported information seeking to better understand the condition: checking the diagnosis and treatment proposed and seeking further information about how to manage the condition. This included wanting to know how to administer the treatment properly. It could also include wanting to be aware of potential side-effects or complications, so that these could be anticipated and subsequently managed, and investigating potential success rates and likely recovery times.

Another issue is the mode of presentation of information and the ways it is expressed. People with different backgrounds (e.g., patients, doctors, nurses) typically describe their information needs in different terms, and they need information to be presented in different forms (Adams and Blandford, 2002). In principle, they also have access to different information resources on the same subject, and, therefore, they may need to negotiate towards shared understanding.

Looking to the future, what needs to evolve is improved information literacy, for both experts and lay people, and new ways of interacting with and around information as the roles of domain experts and lay people co-evolve with the availability and use of new information resources and new ways of interacting with those resources.

3.3   TEMPORAL ASPECTS OF THE SITUATION

Activities with and around information, and human capabilities and system designs, evolve over different timescales:

- Within a single information seeking episode, the needs and understanding of the individual change, as outlined in Section 2.2 above.

- Within any day, different activities are interleaved, typically including a variety of information needs.

- Within any activity, which may last minutes, days, weeks or months, episodes of information seeking are interleaved with various forms of information use. As discussed above (Section 3.1), as more high quality information becomes available "anywhere, any time," the coupling between seeking and use is becoming tighter, and the episodes of seeking are typically shorter, more frequent and more opportunistic.

- Over time, an individual develops both a better understanding of the domain(s) in which they are working and of the nuances of information working.

- Over time, new technologies emerge to support aspects of information work, as well as new information being generated and published.

Most work on information retrieval and even information seeking (with some notable exceptions) focuses on a single information seeking episode. Some technologies have sought to deliver optimal results in a single-shot query; others have supported iteration, but treat the information need as if it remains largely static. Within information seeking, it has been recognised for some time that the need and the information 'consumed' co-evolve (Bates, M., 1989), but the role of people's prior experience, world-view and expertise in the evolving interaction with information is rarely taken into account. For example, Evidence Based Medicine (Sackett, D., 1997) assumes clinicians refer to current best evidence routinely as a one-shot event in the context of clinical encounters. In practice, we found (unsurprisingly) that clinicians rely largely on their established expertise, and that expertise is maintained, or develops, largely through a process of "information grazing" in which understanding is updated through monitoring developments in their specialisms: reading the literature and talking with colleagues. They need to explicitly seek information (i.e., recognise and address a need) when working on the boundaries of their expertise. In these situations, they will typically consult colleagues rather than texts because colleagues better support them in understanding their information needs, and interpreting and validating information.

Carroll and Rosson (1992) discuss the co-evolving nature of systems and uses: that people adopt new systems, then appropriate them to novel uses, and designers may then design new systems that better support both those new uses and other uses that were previously not possible, which lead to further changes in user behavior, resulting on an ongoing, coupled evolution of system design and user behavior (with, of course, concomitant changes in user understanding).

The extended co-evolution of design and use is rarely studied explicitly; rather, snapshots are taken – typically, either of a user study that led to the identification of user needs, leading to the design of a system to address those needs or of a system design that was developed in response to a technological opportunity, and it might have, subsequently, been subjected to evaluation by users (Blandford and Bainbridge, 2009).

In this section, we consider two aspects of extended information interaction over time: the evolving information interactions as an activity progresses and how expertise in information interaction develops over time.

3.3.1   INFORMATION INTERACTIONS WITHIN AN ACTIVITY

Information interactions may be part of many activities, e.g., making a clinical diagnosis, deciding on a holiday location or preparing a writing assignment.

In professional disciplines such as medicine and law, an important role for information is to provide evidence, in the form of precedent, for different kinds of intervention or strategy. In their study of lawyers Kuhlthau and Tama (2001) quote one lawyer who said (p. 30): "The hardest part of the job is figuring out a strategy for a complex case and figuring out what path to take. […] Trying to figure out how it is going to play before a jury." This illustrates both the poorly determined nature of early information needs, as the lawyer is looking for a suitable approach or angle on the case, and also the need to identify information that will suggest potentially successful strategies – in this case, for what to present before a jury.

In medicine, within the UK, there has been a political shift towards clinical audit and Evidence Based Medicine (Adams and Blandford, 2002; Adams et al., 2005b). This demands that clinicians have access to information about current best evidence (e.g., of diagnoses and treatment plans) and best practice. However, compared to lawyers, clinicians have traditionally made little use of such evidence interleaved with their ongoing work, typically relying on a combination of prior training and experience, discussions with colleagues and continuing professional development that is not focused on an individual case. Thus, clinicians have a relatively underdeveloped culture of accessing professional (evidential) information within the context of their day-to-day practice (with some notable, well defined, exceptions such as ward protocols and information on drugs).

A valuable approach to understanding how information interactions relate to and support some broader activity, and also how they can be designed to support them better, is to track them over the course of a work assignment. For example, the work of Kuhlthau, C. (1991), presented in Section 2.2, is based on such a study of students. Her focus was on how the information problem was refined, by the student working out what is possible with the materials available, to reduce the uncertainty inherent in the assignment as presented. Similarly, in a longitudinal study, Vakkari, P. (2001) examined how students' search tactics and relevance assessments evolved over the course of a three month assignment. None of these studies, however, related the full range of information interactions which make up the broader work processes, such as the creation and use of intermediate artefacts during the writing activity.

Conversely, many studies of writing have focused on the text production aspects of writing to the exclusion of other equally important aspects of the process (O'Hara et al., 2002). In a classic study, Flower and Hayes (1981) describe the cognitive processes involved in expository writing. At the highest level of abstraction, they divide the writer's world into the task environment (i.e., topic, intended audience, motivation and the text produced so far), long-term memory and the writing process. The writing process is viewed as the interaction between three broad cognitive sub-processes: planning, translating, and reviewing. Sharples, M. (1996) criticises this view as ignoring the roles of external representations in supporting writing, arguing that these enable the writer to explore different ways of structuring content and to apply systematic transformations, such as prioritising, reversing order, or clustering related items. Other authors highlight the central role that source materials play in supporting writing. Neuwirth and Kaufer (1989), for example, developed a task framework for composing a written document from source materials which consisted of: identifying relevant information in individual source texts, grouping sources by similarities and differences, organising sources into a tree by similarity/difference, and generating document structure by traversing the previously developed hierarchy. This model presumes a one-way flow of activity, where all sources are identified before composition begins.

As described in Section 3.1.2, O'Hara et al. (2002) studied how writers use source documents during everyday writing tasks, focusing on interactions in terms of the material properties of source documents and how these relate to underlying cognitive processes. They reported periods of frequent attentional shift between source documents and composition for quick reference. Again, the focus was on the writing task, apparently assuming that all source documents were available at the start of the process.

In order to understand the writing process from a more holistic perspective, Attfield and Dowell (2003) conducted a study of journalists writing new reports that identified a tight and iterative integration of information seeking and information use in practice and the evolution and use of constructed resources (see Section 2.5). Attfield et al. (2008b) studied a single music journalist completing a feature assignment and similarly found integration of multiple activities and the evolution of a resource space. Periods of information gathering and use were interleaved. Gathering was typically in support of ideas generation, to be followed by periods of consolidation, where intermediate artefacts (such as a proposal, an interview plan or an outline article) were generated, evaluated and used as a starting point for the next phase of activity. Recognising the cyclic nature of creative work, Shneiderman, B. (2000) proposes a four-phase framework outlining activities which he argues are key to creativity, namely collect, relate, create, and donate. This highlights again the tight coupling between information finding and use, as creating and collecting cannot be done independently of each other.

In summary, information interactions are closely interleaved with other actions within most knowledge-based activities, but they have rarely been studied explicitly within the study of the broader activity. The information journey presented in Chapter 4 redresses the balance, locating information interaction within the broader, evolving activity.

3.3.2 DEVELOPING EXPERTISE IN INFORMATION INTERACTION

In an earlier Section (3.2.2), we briefly discussed the importance of communities of practice in empowering people to develop a range of behaviors. A key set of skills centre around information interaction. Various aspects of this have been investigated in the past: the necessity to learn to use particular information resources, how to formulate a search query, how to refine queries over time, how to assess the relevance of results; and how skills and knowledge evolve over time.

Familiarity with particular resources is an often overlooked aspect of expertise. This applies to both physical and digital resources: in the physical library, it is necessary to understand not just the general principles of library organisation but also the specifics of how a particular library "works"; the same is equally true for digital libraries. As discussed above (Section 3.1.1), people raised in different cultures can have great difficulty in learning to use Western libraries. Makri et al. (2007) studied the understanding that people develop of particular information resources, and they found that people draw on analogies from other domains (not just physical libraries, but general search engines) to understand how digital libraries work, and that features such as access control mechanisms can prove to be serious blocks to understanding and use. Blandford et al. (2001) found that users invariably chose to work with familiar resources when given a choice. Familiarity with a particular DL includes many facets: its structure, features, contents, effective indexing terms, etc. This ability to probe and gain familiarity is an important aspect of expertise.

For experimental purposes, most researchers (e.g., Hsieh-Yee, I. (1993); Sutcliffe and Ennis (2000)) treat expertise as a binary state, either novice or expert. Two key variables that have been studied are information seeking expertise and subject knowledge.

Some researchers (e.g., Smith et al. (1989)) have focused on the importance of domain knowledge as a component of expertise. Vakkari, P. (2001) studied development within one extended searching episode (that took place over several weeks). He reported that users' searches became more focused as the nature of the information problem becomes better understood. As the search progressed, so the problem definition and query terms became more clearly defined, and users were more easily able to make relevance judgements. In this way, we see a development of searching with respect to a particular information problem. Vakkari argues that the effects that are observed are primarily due to development in subject expertise and not in searching expertise.

Hsieh-Yee, I. (1993) found that information seeking experts tend to explore synonyms, to establish what effects these have on search results, whereas novices (even subject experts) do not. Within their subject area, the differences were relatively small, but outside their area differences were much greater: experts were able to use on-line tools such as thesauri to assist in generating alternative search terms whereas novices relied on their own intuition in selecting terms. Stelmaszewska and Blandford (2002) found that novice users have very little persistence in query formulation, usually giving up a particular search after only two or three attempts. In contrast, Fields et al. (2004) found that expert users – in their case, librarians acting as information intermediaries for users who had reasonably well defined information requirements – have well honed strategies for refining search terms to achieve a results set of the desired size and quality. For example, an expert may explore synonyms and probe results to see which terms are working well, whether there are 'distractor terms' that need to be explicitly excluded from the search terms and whether there are alternative terms that appear in documents that might usefully form part of the search query.

Once the user has received results, their relevance to the task at hand needs to be assessed. As discussed above, Stelmaszewska and Blandford (2004) found that users in physical libraries have a range of strategies for assessing the relevance of documents, most of which do not currently have digital surrogates. With a limited window onto a search results list, it appears that users rarely scroll through more than the first page or two of results. An obvious consequence of this is that users rarely even see results that are not returned near the top of the list, highlighting the importance of having effective ranking algorithms that return the most user-relevant results near the top of the visible list.

As discussed above, researchers such as Hsieh-Yee, I. (1993) have used two dimensions of expertise, in information seeking and the subject domain. Cothey (2002) criticises studies such as that of Hsieh-Yee, I. (1993) as confusing expertise with experience, assuming that sophistication of searching is correlated to the amount of searching done. In a longitudinal study of changing search practices by web users over a ten-month period, Cothey found that, rather than becoming perceptibly more sophisticated, users' behavior was more appropriately characterised as shifting from searching to browsing and focusing on fewer selected sites – both features of information gathering that indicate streamlining of search effort. Thus, Cothey appears to be arguing that familiarisation with particular resources, appropriate to their usual needs, is an important element of user experience. However, another aspect of her argument is that most users do not appear to develop the sophisticated information working skills observed by Fields et al. (2004) in their study of librarians; people only develop such skills if they have compelling reasons to do so. This is consistent with the findings of Warwick et al. (2009), who studied the information practices of information management students over a two and a half year period. They found that participants generally chose to use generic search engines such as Google in preference to more sophisticated and selective information sources. They would continue to use well practiced searching strategies as far as possible, completing tasks with minimum information seeking effort. New ways of discovering and selecting information were only adopted when immediately relevant to the task at hand, and assignments were generally chosen or interpreted in ways that minimised the need to develop new strategies. For most people, there is little incentive to become more of an information interaction expert than is strictly necessary for the tasks at hand.

Information interactions evolve in response to a range of contextual factors, and yet there are commonalities in how information interactions relate to other aspects of people's activities; this is the topic for the next chapter.

CHAPTER 4

The Behaviors: Understanding the "Information Journey"

In the previous chapter, we have presented various viewpoints on how people interact with information in a situated way, highlighting a range of information behaviors and their evolution over time.

In this chapter, we present an "Information Journey" model for reasoning about the design of information interactions. Like any model, the Information Journey highlights some features of information interaction while downplaying others. It is a framework that supports the comparison of findings from different studies, and that also supports reasoning about design, a topic that we turn to in the following chapter.

4.1 THE DERIVATION OF THE INFORMATION JOURNEY

The Information Journey model is derived from findings from a series of studies conducted in a variety of settings over several years. These studies have generally sought a "bottom up" understanding of what people really do and how information integrates with their professional and personal lives. A few of the studies have involved controlled experiments, most of those focusing on novel system designs (e.g., Stelmaszewska et al. (2005); Attfield et al. (2008a)) and their effects on how people interact with information. Necessarily, these studies have been "in the small," focusing on local effects rather than the broader situations within which natural information work takes place. The majority of studies have been qualitative, gathering rich data from people, studying how they interact with and use information, through both technical and social systems, in their broader working context. It is these situated studies that have directly informed the development of the "Information Journey."

Data has been collected from a variety of settings, using techniques including observations, in-depth interviews, Contextual Inquiry (Beyer and Holtzblatt, 1998), focus groups and system logs. The different studies have had various detailed focuses, from how engineers find information in a digital library to how lawyers make sense of information in a large fraud investigation or students prepare a piece of coursework.

Data gathering and analysis has adopted different starting points and frameworks. Sometimes, only broad themes and questions were considered, and a Grounded Theory (Strauss and Corbin, 1998; Charmaz, C., 2006) approach was taken, interleaving data gathering and analysis, evolving a grounded model of information interaction in context. At other times, a particular model or theory (such as Cognitive Systems Engineering (Rasmussen et al., 1994), Ellis' (Ellis and Haugan, 1997)

Information Seeking Model or Communities of Practice (Wenger, E., 1999)) has been applied – either prior to data gathering or as a 'lens' through which to analyse the data. The aim in all studies has not been simply to understand information work but to develop theory that can inform the design and deployment of future technologies.

Details of individual studies, including the participants, the settings, the methods and the focus of investigation are presented in more detailed papers; the synthesis presented here represents an abstraction over several studies. The initial "information journey" was presented by Adams and Blandford (2005); what we present below is a development from that work, extended to a range of settings.

4.2   THE "INFORMATION JOURNEY" FRAMEWORK

The information journey encapsulates phases of:

- Recognising an information need (also called an "anomalous state of knowledge" (Belkin et al., 1982a)).

- Acquiring information (possibly through active searching, or maybe by serendipitous finding or being told).

- Interpreting, and often validating, that information.

- Using the interpretation (e.g., in writing or decision making).

These phases evolve with the activity, resulting in an evolution of understanding and interaction (Figure 4.1). The phases are not necessarily sequential; for example, information may be acquired incidentally (without the individual having previously recognised the need), and it may be necessary to find and interpret (or make sense of) a lot of information before any of it is overtly used.

To illustrate the information journey and how it is instantiated in a variety of contexts, we provide brief accounts of the information journey of patients in healthcare and of journalists writing news articles. We then draw out themes based on particular configurations of the information journey: sensemaking, information encountering and serendipity, needs engendering further needs, and the role of anticipation in guiding information interactions.

4.2.1   EXAMPLE: A PATIENT'S INFORMATION JOURNEY

Our first example draws on material that has been discussed earlier in this lecture, on how patients find and use information.

Recognise need: People's need for health information may be initiated by a recognised problem (symptom) or by an external event (e.g., a news article on MMR vaccine may make parents want to know more about the issues before getting their children vaccinated). As discussed above (Section 3.2.3), much health information seeking is centred around the clinical encounter, with a view to

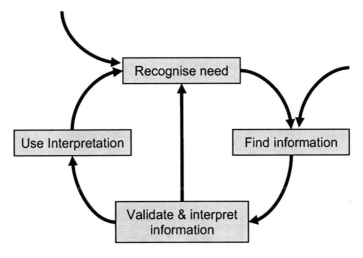

Figure 4.1: The basic information journey.

answering questions such as "what will the doctor want to know from me?" and "what are the likely side effects of this treatment?"

Find information: The ease of acquiring health information depends on many factors. For example: prior to the clinical encounter, many patients may not have the appropriate vocabulary to be able to generate effective search terms; also, online health information is currently organised in 'silos', and finding suitable information often depends on being able to identify appropriate 'silos' in which to search.

Validate and interpret information: Interpretation of information also often poses a challenge: interpretation refers not only to the ability to understand the words, but also on the ability to contextualise them to the current situation, and derive meaning that can inform use. Particularly in the healthcare arena, challenges to interpretation include a consideration of who the information is written for: not only is information used in different ways by different user groups (general practitioners, nurses, consultants, patients, etc.), but it is also written with particular assumptions about prior knowledge and the purpose for which the information might be required. Validation of information can also pose a challenge; many lay readers of health information have limited skills in checking the validity and appropriateness of information – e.g., whether it is actually advertising, promoting the products of a particular supplier, or whether it is merely the opinions of another lay person (Sillince et al., 2004).

Use interpretation: Finally, in the clinical context, use typically involves making decisions – e.g., about whether to consult a doctor – though there are situations where the concern is simply with 'making sense' of the situation, with no direct 'use' outcomes.

In many situations, the information need evolves over time, as the patient's understanding of their condition grows, as the questions change in response to developments in the condition, and in response to interactions with health professionals.

4.2.2   EXAMPLE: A JOURNALIST'S INFORMATION JOURNEY

Our second example also draws on material discussed earlier in this lecture, namely the work of journalists, information professionals for whom information interaction is an essential part of their daily work.

Recognise need: For many newspaper journalists (noting that there are many different roles within the newsroom), the information need is typically initiated by the requirement to have information to support a particular interpretation or 'angle' taken on a recent event (Attfield and Dowell, 2003). For example, many new stories relate a current event (e.g., a train crash) to previous similar events (e.g., earlier train crashes with comparable attributes, or earlier tragedies in the same location) to support the reader in making a particular kind of sense of the story (e.g., the largest train crash since a previous infamous incident).

Find information: The process of finding information to support the story is often exploratory. A variety of resources are used, including specialist news archives and general web resources. In determining what information is likely to be relevant, journalists are working with constraints, such as:

* The angle: the story that is being developed relative to the known facts.

* Newsworthiness: the likely interest of readers of the story, and

* Accuracy of information.

These relevance assessments are often made in anticipation of how the information will be subsequently interpreted and used, a topic to which we return in Section 4.6.

Validate and interpret information: Facts need to be checked. Interpretation in this context often includes triangulating information from multiple sources. In particular, the "angle" of a story can guide interpretation of information. Attfield, S. (2005) cites an example of an angle that evolved during the events of nine-eleven:

> " ...I remember on the day [September 11th, 2001] that by the time of the second plane, I and others were saying: 'This must be an act of terrorism, because this is not coincidental, an accident....' So had I been writing the story, I would have begun building up information to support my hypothesis that the acts of September 11 were terrorism.

The standard journalistic questions of who, what, why, when, how would have been asked about the events against the backdrop of my hypothesis of terrorism" (pp. 110-111).

Due to inherent uncertainty in the situation, with new facts coming to light potentially challenging an interpretation, editors changing their minds, as well as considerations such as what competing journalists might be writing, assessment of the relevance of any information is frequently provisional. Consequently, journalists often need to be able to interact fluidly with information including refinding information that was initially judged to be of low relevance as requirements evolve; there may be a cycle of identifying needs, finding and interpreting information that repeats before the information is used.

Use interpretation: Finally, use supports the writing of news articles.

Depending on the nature of the article being written, the entire information interaction may be over quickly, as the article is written to a tight deadline, or there may be a need to gather more information (e.g., in response to new information about the event).

4.3   MAKING SENSE OF INFORMATION

The two examples above illustrate the simple application of the Information Journey framework. In practice, many information interactions involve adaptations of the framework. One of the most widely studied approaches to interacting with information is sensemaking, as discussed in Section 2.3. This sensemaking loop is illustrated in terms of the Information Journey in Figure 4.2. In this case, the eventual use of information is downplayed relative to the finding and interpreting of that

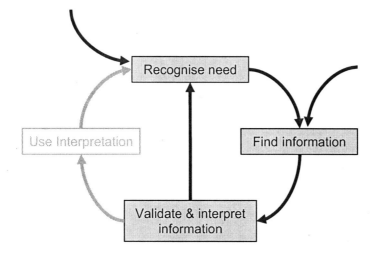

Figure 4.2: The basic sensemaking loop.

information. In terms of the foraging and sensemaking loops of Pirolli and Card (2005), information finding corresponds to foraging, while interpretation corresponds to sensemaking, and the ongoing identification of new information needs mediates between sensemaking and foraging activities.

In this case, we illustrate the Information Journey through an example taken from our work with lawyers. This demonstrates the interplay between top-down and bottom-up processes as this plays out during sensemaking.

When the legal conduct of a company is brought into question, the concerns raised can trigger an investigation on behalf of a regulatory authority, or as a prelude to possible litigation. Such investigations, carried out by teams of lawyers, typically involve the recovery and analysis of large quantities of electronic evidence such as emails, spreadsheets and word-processor documents (a process referred to as e-discovery).

Given the typically vast size of the document collections involved in e-discovery (e-discovery requests for email alone can result in tens of millions of documents), lawyers face a significant sensemaking challenge. They need to use this information to form an understanding of events and actions within the company that have a bearing on the investigation.

Given the amount of information involved, one important activity is to narrow in on areas of importance. Based on case studies of three e-discovery investigations, we found that this narrowing is achieved through two complementary and reciprocal forms of focusing: data focusing and issue focusing (Attfield and Blandford, submitted). Data focusing involves identifying, extracting and structuring information relevant to a stated set of investigation issues; here issues involve theories and questions arising from those theories. Issue focusing, in contrast, involves refining the issues under investigation, for example, through the examination of information and the formation of new interpretations. Data focusing is a bottom-up process; issue focusing is a top-down process and gives rise to new, more focused questions.

Both kinds of focusing are essential to sensemaking, but whilst data focusing is a frequent subject of study (possibly because it features explicit actions, e.g., search), issue focusing is less well understood.

In the investigations we studied, issue focusing occurred at many levels, from the articulation of major areas of enquiry to the identification of small-scale, local questions. We illustrate issue focusing with a small-scale example. By analysing documents and witness interviews, the investigators constructed explicit narratives of events (in the form of chronologies), each relevant to a particular area of investigation. Initially, knowledge of an event, such as a meeting between protagonists, might be cued by the discovery of an email exchange. A single email, however, would only provide partial and potentially inconclusive evidence. But once the event was revealed by this evidence, the lawyers would want to know more about it: what was said, who was there, what the outcomes were, etc. Indeed, the meeting may not have taken place at all, but it may have been replaced by a telephone call. Once investigators became aware that an event might have occurred, it was important to find out more about it. Hence, new information and its interpretation resulted in the production of new, more focused information needs, which then led to new information seeking, and so on (as illustrated in Figure 4.2).

4.4   INFORMATION ENCOUNTERING AND SERENDIPITY

Sometimes, information is encountered and used without an explicit need ever having been identified. For example, Attfield and Blandford discuss the use of current awareness services that provide information on legal developments in the general area of a lawyer's interests. Perhaps the most interesting of such encounters are generally regarded as "serendipitous."

There is a widespread understanding of "serendipity" as being a lucky discovery that is not planned. Serendipity has been studied in three distinct areas: science, chance meetings and information seeking. In science (e.g., Kubinyi, H. (1999)), the term refers to accidental discoveries – e.g., of new processes or theories. In meetings and information seeking, serendipity refers to chance encounters that are, in some way, fortunate.

In the information seeking literature, there is an assumption that the individual is actively engaged in information seeking when serendipitous information finding occurs. For example, Erdelez, S. (2004) studied "information encountering," which she describes as a particular form of serendipity in which the individual is looking for information on one topic and encounters information on a different topic that is also of interest, precluding the idea that information might be encountered outside the context of a deliberate information interaction. Similarly, Toms, E. (2000) discusses three modes of information seeking: about a well-defined topic, for information that cannot be clearly articulated but will be recognized when found, and through serendipity.

Foster and Ford (2003) present several related but different views of what serendipity is, including finding items that are similar to that already found; the accidental discoveries of science; and the outcome of a productive, systematic search strategy. They note that "accounts of the creative process of research do not leave serendipity as Walpole's classic 'fortuitous discovery,' but hint at something more active, operating at the edge of consciousness" (p. 323). In this, they highlight the importance of the prior skills, knowledge and attitude of the individual as well as the design of the information provision systems in creating the conditions for serendipity. In terms of impact, they note two main outcomes: addressing the existing problem or taking the researcher in a new direction. They include examples of chance encounters without active seeking, such as people hearing an interesting article on the radio.

In our studies of humanities scholars, we found that many of their research projects arose out of serendipitous encounters with information that resulted in information being validated and used directly. As noted above (Section 3.1.1), the library is the humanities scholar's laboratory, in which texts are discovered and studied. Texts, as the material of study, are often valued as a personal resource, more valuable if they are *not* widely accessible. A discovered text may cause a scholar to develop a new research agenda based on an unexpected interpretation they can now form, at least tentatively. This may suggest a new goal (or target use), leading to new things to find and interpret etc. (Figure 4.3).

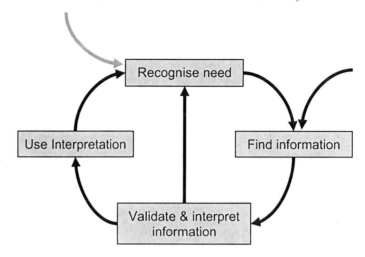

Figure 4.3: The core elements of a discovery-based information journey.

Naturally, as discussed in the next Section 4.5, other information will usually be required to support the research, leading to a more traditional Information Journey, but the journey can start with an unexpected encounter, rather than an explicit need.

4.5   WHEELS WITHIN WHEELS: THE RECURSIVE NATURE OF MUCH INFORMATION WORK

We have described the Information Journey as an evolving cycle of information interactions as if each new interaction built directly on earlier ones. This view is also implicit in descriptions of information seeking as "foraging" (Pirolli and Card, 1995), "berrypicking" (Bates, M., 1989), or "information grazing" as described in the Introduction (Chapter 1). In practice, and as is recognised in many of the descriptions of sensemaking, the need to understand a situation may often spawn many more needs to understand different elements of that situation, which may in turn spawn further needs. This was the process at work in the legal investigations discussed above.

Newman et al. (forthcoming) present a graphic example of this in their discussion of a researcher (Bella) investigating Isabel de Mortimer's marital status:

"Genealogical data in a highly respected peerage (Cokayne 1887), indicated that de Mortimer had married again in 1275, to Ralph de Ardern. But Bella became increasingly confused as her researches revealed de Mortimer continuing for a further ten years to behave in the independent manner typical of medieval widows. She eventually realised that she must check the original manuscript in which the marriage to de Ardern was recorded."

This is an example of information that does not "make sense," creating a locus of activity around the "information interpretation" stage of the Journey. It does not make sense because the information does not fit a considered interpretation in an unambiguous way. This led, in turn, to further information needs to resolve the growing ambiguity in what was known. (It was eventually established that there were two women of the same name whose histories had previously been intertwined into one narrative.)

The emergence of new information needs, whether sub-topics or related, is characteristic of many information interactions.

4.6   ANTICIPATING FUTURE DEMANDS IN THE INFORMATION JOURNEY

The basic information journey has been presented as if the flow of activities is in one direction (from needing to finding to interpreting to using). This is not always the case: people sometimes anticipate the demands of future stages and use that anticipation to guide current activities.

In the discussion of journalists, we noted that information finding (and particularly relevance assessments) is often conducted with an eye to the future use of that information in writing an article. This is a situation in which there is interdependence between use and need. Attfield et al. (2003) used the idea of writing as a type of design activity (Sharples, M., 1996) as a point of departure for understanding information need uncertainty in relation to the task as a whole. Design problems are frequently radically under-specified and uncertain, with this uncertainty resolved through iterations of analysis and synthesis (Lawson, B., 1997; Schön, D., 1983). Attfield, Blandford and Dowell noted that where information seeking is embedded within a wider task, a reciprocal relationship occurs such that information seeking is shaped by the needs of the task, and yet the evolving task is shaped by the information found.

A similar situation exists for students completing course assignments; Kuhlthau, C. (1991) reported that students experience early uncertainty as they find out what information is available to address the problem, and they interpreted the observed problem refinement as reducing uncertainty as the student establishes what is possible with the information available (reciprocity between need and finding). We conducted a study of students completing course assignments that showed that the anticipation of future demands goes further: that students also anticipated the demands of interpretation and use, and that this anticipation guided their selection of information resources in which to find information in the first place.

This particular account is part of a longitudinal study in which we sought to establish how Information Management students developed information working skills (Warwick et al., 2009). Our initial assumption was that their information seeking skills would develop significantly over the period of our study (which took place over 2.5 years). Cothey (2002) earlier found that students typically became more selective in their information sources over time. One of our key findings was that students often redefine the information need in order to minimise the demands of interpretation and use. The information need is defined by the assignment description, which is an "imposed problem," rather than an "owned problem," in the sense that it is defined by an external party, and many students focus on the question of how to deliver a desired solution to the problem that matches the problem owner's requirements rather than their own. Most students are both risk-averse (preferring to answer the assignment in a safe way) and strategic in how they apply their effort.

What we found was that, anticipating the demands of interpretation and validation, students typically select sources in which they have confidence while also selecting topics (or reinterpreting the question) so as to minimise the interpretation demands (Figure 4.4).

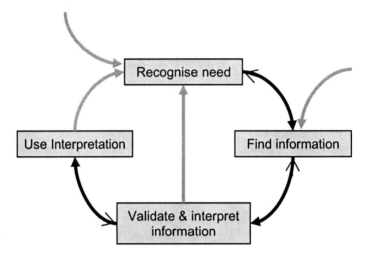

Figure 4.4: The core elements of an anticipation-based information journey (with thinner arrows indicating feedback anticipating demands).

This is one approach to minimising effort (and also minimising risk). Whereas prior work on how people minimise effort, or satisfice, in information seeking (e.g., Prabha et al. (2007)) has focused on the stopping condition, this work showed that people can also redefine the need to reduce the demands of information seeking and interpreting. Whereas most prior work has assumed an immutable information need, whereby people determine when they have sufficient information to address their need, this work highlighted a reciprocity (or interdependence) between the need, the validation and the use of that information.

4.7   SUMMARY

In this chapter, we have presented and exemplified a view of information interaction as an evolving "information journey." This is a high level description of the various phases of information interaction that have most commonly been studied in isolation – for example, as information seeking, or writing, or knowledge management. The information journey view brings these different phases of activity together; as we show in the next chapter, this more contextualised view can be used as a framework for considering new kinds of design solutions that enable different phases of activity to be linked together to create smoother overall information interactions.

CHAPTER 5

The Technologies: Supporting the Information Journey

In the previous two chapters, we have outlined the situated aspects and the process of people's interactions with information. Implicitly, these interactions have taken place with both physical and digital artefacts which were designed, or which simply came together in configurations that were part designed (whether by information or technology specialists or by the individual) and part evolved, to meet an individual's needs. For example, a traditional research library is designed with carrels, which people adapt to their individual needs by gathering together books, papers, etc., to support their work. Similarly, a digital library is organised in a particular way for the purposes of browsing, typically augmented by a search facility, but an individual may create their own personal "library," or research space, by storing some documents locally on their computer or making them easily retrievable by using a bookmarking or "bookshelf" facility, or by printing out material and spreading it around the workspace. They may also carry materials around, annotate, share and reorganise them.

In these ways, the information practices and tools used by individuals co-evolve, as people appropriate existing tools and make them work in new ways, and designers develop new tools to support both existing practices and anticipated future needs.

The 'information journey' is intentionally inclusive. It is intended to present a more complete picture of information interaction than is usually on offer in the literature. As such it is intended to cast light on aspects of our experiences with information which may previously have been obscured. A good deal of prior research which can be said to come within the ambit of information interaction, and also the tools that have been designed to support information interaction, have in fact limited their focus to the active seeking of information. This is to the exclusion of phenomena such as passive information encounters and sensemaking, which have received far less attention and yet are significant components of the situated information journey.

In this chapter, we review a selection of existing and prototype digital tools intended to support a range of information interactions, and the roles that they serve in supporting the information journey. We focus in particular on tools for supporting information encountering, sensemaking and the integration of finding and use.

5.1 DESIGNING TO SUPPORT INFORMATION ENCOUNTERING

The significance of chance encounters with interesting information has been recognised, whether while looking for other information (per the definition of encountering proposed by Erdelez, S. (2004)) or while doing other things. Whereas physical libraries support some forms of serendipity through the physical organisation of material, and the fact that the library user has to pass that material while looking for a particular resource, it is necessary to find new ways to design for such encounters with information in the digital world. Two approaches that are being investigated are designing for serendipity and awareness mechanisms.

5.1.1 DESIGNING FOR SERENDIPITY

There have been a few attempts to design to support serendipity in information seeking. Many of these systems support what Erdelez, S. (2004) calls "encountering," in that they presuppose that people are already seeking some information, and the challenge is to present other information that might also be of interest.

Toms, E. (2000) proposes four means of facilitating serendipitous interactions: random information generation, matching user profiles, identifying anomalies via poor similarity measures, and reasoning by analogy. Toms and McCay-Peet (2009) have implemented and tested a system proposing suggested pages based on using the first paragraph of the current page as a search query; in a laboratory study, they found that 40% of their users followed suggestions; however, as they note, users are likely to behave differently in natural search settings.

Following a similar approach, Campos and Figueiredo (2002) present a system, Max, that supports "the problem of sagacity" (p. 52). The approach involves presenting novel information from the web to the user, based on "little jumps [to] adjacent concepts on the conceptual map that surround the user's interests" (p.58). A preliminary evaluation of the system showed that about 10% of the suggestions made by Max were regarded as very valuable by participants, leading the authors to conclude that the approach is highly promising.

5.1.2 AWARENESS MECHANISMS

While systems to support serendipity have, to date, been implemented in the context of searching for information (more or less closely related), others have investigated the development of background awareness mechanisms that enable people to be aware of information that they were not explicitly seeking.

Examples include systems to support "just in time information retrieval," which track people's activity and perform background searches on queries related to the current activity. An example of such a system is PIRA (Twidale et al., 2008), described below (Section 5.3.2).

Farooq et al. (2008) present an awareness mechanism that enables people to be alerted to relevant developments in their research area. They implemented and tested three RSS feeds based on CiteSeer data. One notified the user when their papers were cited, the second when papers on their area of interest (based on keywords) were published, and the third when papers related to their work (e.g., through co-citation) were published. Three alternative presentations of information were tested: reference; reference and abstract; reference and context (where 'context' showed the text citing the paper of interest, or the keywords in common). In a formative evaluation study, they found that in the first instance (notification of papers that cite one's own), people preferred to see the reference and the context in which one's own work was cited, whereas in the other two situations reference and abstract was considered most useful. The common theme was that information about new papers in the area was considered most valuable if enough contextual information was provided to form an immediate judgement on whether the paper referenced is of significant interest. In a subsequent longitudinal study, in which participants were asked to work collaboratively on a writing task, three feeds were sent to participants in the course of the task; participants' views of the feeds were generally positive, but participants expressed the view that such a facility would be more valuable if it were available for their personal areas of interest (rather than the imposed problem that the study task represented). Overall, this work suggests that such an awareness mechanism has potential value, but it has not yet evaluated how the awareness mechanism could fit into people's ordinary work activities.

Adams et al. (2005a) describe the development of an organisational awareness server that presents information about activities, opportunities and news in the organisation. This system displayed information whenever it had been idle for a few minutes, drawing on a database of information items that had been supplied by other people within the organisation. Adams et al. identified two key features of the system design that contributed to its success; the first was that many stakeholders became involved in the system design, and they felt a sense of "ownership" of it, so that the information that was displayed was perceived as having real value to people in the organisation (and not just to the people selecting the information for display); the second was that the awareness server was activated when the computer was idle, which was typically at times when people were taking a break to have refreshments or talk with colleagues, and they were more receptive to information that was not directly pertinent to the current task.

5.1.3   INFORMATION ENCOUNTERING: SUMMARY

Extant approaches to supporting information encountering take one of two approaches to selecting information to display – either semantically related to ongoing activity or effectively random – and one of two approaches to when to present information: among results of an ongoing search or as background information, to be noted during "quiet" times. What is clear is that information encountering can be valuable, if the information is of interest, but that it should not be a distraction from ongoing activities, and that there remain design challenges in anticipating what might be of interest and alerting people to that information in unobtrusive ways.

5.2 DESIGNING TO SUPPORT SENSEMAKING

In terms of technology development and deployment, much more effort has been invested in supporting the provision of information (e.g., through digital libraries, databases and the Web) and finding of information (primarily through search engines and, to a lesser extent, through information architecture) than in the interpretation of information.

People often make sense of information through somebody else's interpretation. Thus, as discussed in Chapter 4, lawyers go to commentaries before they access legislation, while students often choose text books rather than more challenging research papers. Journalists provide us with off-the-shelf interpretations which we can choose to accept or not. We think of this as sensemaking-by-proxy, and, sometimes, it is enough; however, often, it is not.

In this section, we consider technologies, both textual and visual, for supporting sensemaking. We briefly describe Search Friend, a set of interfaces that support different levels of exploratory search by presenting the information seeker with different amounts of contextual information about search results; we discuss visual analytics, a growing field of investigation on visualisation approaches to supporting sensemaking; and, finally, we discuss the role of spatial hypertext systems in supporting the organisation of information for sensemaking.

5.2.1 TEXTUAL SUPPORT FOR SENSEMAKING

There are many possible ways of supporting sensemaking; as discussed above, in simple cases, addressing an appropriate audience in writing may be sufficient. While a user is in the early stages of making sense of an area or even making sense of what it is they might like to know, it may be necessary to support exploration and give access to alternative ways of describing the same phenomena.

We have been investigating one approach to supporting exploratory search (White and Roth, 2009) through the "Search Friend" series of interfaces (Diriye et al., 2009). Three interfaces, identical in most respects, were developed, with different levels of contextual support for users showing how suggested terms (i.e., terms that have been identified as relating to the user's search terms) are distributed and used in different web pages. A pilot study found that a more informative approach to presenting suggested terms in context provided better support for sensemaking (when users had little prior understanding of the topic to be investigated, and when the topic is rich). However, a simpler interface was preferable for known-item searches, because the richer interface was inappropriately distracting. The Search Friend interfaces – particularly Search Friend II, which provides contextual information as shown in Figure 5.1 – highlight the importance of context for making sense of digital resources.

5.2.2 VISUAL ANALYTICS

"Search Friend" focuses on the words, and making sense of the topic by facilitating exploration of it; visual analytics is emerging as a research area in which people interact with graphical representations

Figure 5.1: An example Search Friend II interface, developed and tested by Abdigani Diriye, showing search terms and suggested terms in context. Reproduced with permission.

of information in order to make sense of a situation. Two examples of interactive visualisations for making sense of information are Jigsaw (Stasko et al., 2008) and KDVis (Faisal et al., 2007).

Jigsaw (Stasko et al., 2008) is an interactive visualisation to support investigative analysts making sense of collections of documents that comprise evidence in an investigation. The design is based on work such as that of Pirolli and Card (2005), supporting both sensemaking and information finding (see Section 2.3). Jigsaw uses information extraction to identify entities in raw texts and then show connections between entities across documents. Users can then view chronologies, relationship diagrams and groupings of information. Jigsaw has been evaluated by being used in contests such a Visual Analytics Science and Technology (VAST), although the focus to date has been on technical developments.

KDVis (Faisal et al., 2007) exploits visualisations of concepts and relationships. It was developed to support researchers exploring and interacting with their literature in order to make sense of the literature domain. For example, by selecting an author in the top-left window (Figure 5.2), the user can see more about that author's publications (top right), about who has cited that author (bottom left) and about citation relationships between publications (bottom right). The design of

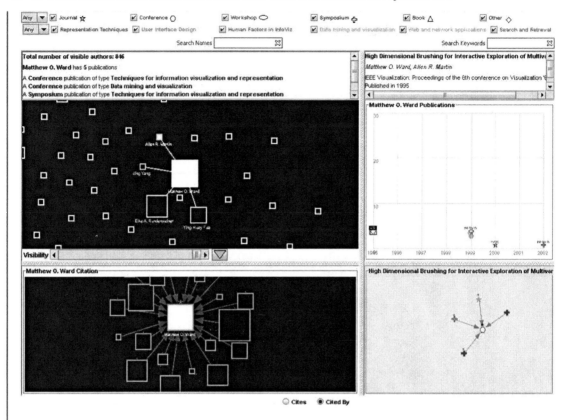

Figure 5.2: The KDVis interface, developed and tested by Sarah Faisal, showing information about authors and papers. Reproduced with permission.

KDVis was based on a requirements study (Faisal et al., 2006), in which people were interviewed to find out how they conceptualise the literature domain, so the implementation matches people's conceptual structures of the domain as closely as possible. A marking feature that people could use in any way they wished was also implemented, and was found to be valued as people used it in personalised ways. A qualitative evaluation study of the tool showed that designing for interactivity, conceptual fit and appropriation are all important features that support people in making sense of the data that is presented.

5.2.3   SPATIAL HYPERTEXT

Sensemaking can involve exploring different ways of structuring information in order to get a sense of the whole from a series of parts. Spatial hypertext systems support this kind of sensemaking using flexible visual layout to indicate connections between material. Here, we present Garnet as an example of a spatial hypertext system that integrates functions for both finding and organising documents.

Exploiting both text and visual layout, Garnet (Buchanan et al., 2004) enables people to find and organise documents to create individualised semantic structures and support information organisation (see Figure 5.3). Garnet integrates spatial hypertext facilities for organising documents

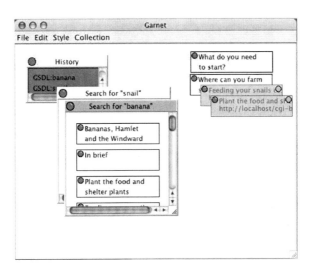

Figure 5.3: The Garnet interface, developed and tested by George Buchanan, which supports users in gathering and restructuring information to support sensemaking. Reproduced with permission.

with digital library access for finding new documents. The user can organise documents manually through spatial layout and can "scatter" a group of documents to cluster them with other documents that are most similar. In these ways, the user can iteratively make sense of a corpus of documents by interleaving information seeking and information structuring. Preliminary evaluation of Garnet has shown that people can integrate these features fluidly and effectively.

5.2.4   SUPPORTING SENSEMAKING: SUMMARY

Tools to support sensemaking are emerging, but few are mature. Sensemaking demands depend on both the information problem and the person's prior understanding. Much support for sensemaking necessarily involves other people. Early tools, both textual and graphical, that support people in understanding the context of information or in integrating information across resources are showing promising results.

5.3   DESIGNING TO INTEGRATE INFORMATION SEEKING AND WRITING

Garnet (discussed in Section 5.2.3) incorporates information seeking with information structuring. Information structuring is most commonly a precursor to use, most typically in writing. Here it is presumed that human sensemaking acts as a 'backdrop' or 'glue' mediating between the information need formulation and the way that information is ultimately put to use. We present two brief examples: background information seeking during writing and retrieval of previously found information during writing.

5.3.1   BACKGROUND INFORMATION SEEKING DURING WRITING

To illustrate possibilities for integrating information finding within an ongoing information activity, Twidale et al. (2008) present the Personal Information Retrieval Assistant (PIRA). PIRA is a writing tool that incorporates just-in-time information retrieval, supported by ambient search (in which queries are formulated based on the current writing activity). Twidale et al. argue that writing and searching should be interleaved, and that the process of writing helps the author to better understand the writing task and hence the information needs of the task. They have followed the same approach in the development and evaluation of PIRA: deploying early prototypes to learn more about the design problem and user needs. The result has been a series of pilot studies and gradually evolving prototypes that incrementally address needs established from earlier studies.

5.3.2   RE-FINDING INFORMATION DURING WRITING

Whereas PIRA focuses on finding new information, our final example in this section supports the re-finding of information that has already been discovered. It is based on work by Attfield et al. (2008a), studying the activities of newspaper journalists in the news room (as discussed in Section 4.2.2). In order to support sensemaking-by-proxy for their readers, journalists use a good deal of background news information to locate events in a context. Consequently, they need to quickly gather multiple sources of information from digital news archives prior to writing. As the task progresses, they then need to interact with these sources frequently and fluidly. They may need to revisit documents many times as they read and extract different kinds of information. The thing that is difficult for them is to predict in advance what information is going to be important when it is initially encountered. The task is uncertain and evolving in the journalist's mind and there are also external sources of change (e.g., an editor changing his/her mind about an angle). NewsHarvester is designed to support low-cost, fluid interaction with gathered documents in the context of this kind of task evolution and uncertainty.

A search component (shown on the left in Figure 5.4) supports the journalist in performing keyword searching; a selected document can be displayed (centre pane), and the text composition area (shown on the right) can incorporate text selected from other documents, and it retains a link back to the full text article, so that the article can be easily retrieved again for review or reuse.

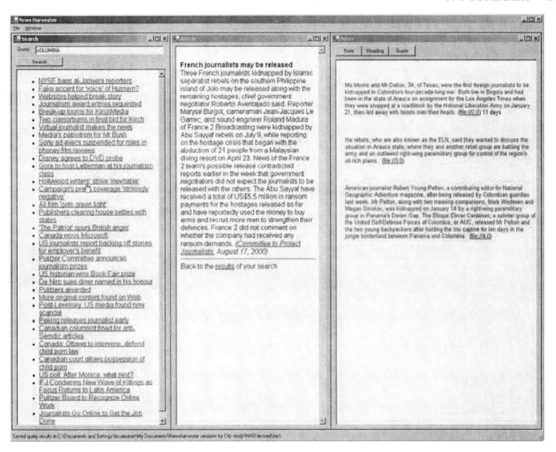

Figure 5.4: The NewsHarvester interface, developed and tested by Simon Attfield, which supports users in linking information being built into today's story back to its source.

Attfield et al. (2008a) report a user evaluation conducted with journalism students in which the system was compared with printing and standard drag-and-drop (without a link) as alternative (and commonly used) information gathering strategies. The results strongly favoured the NewsHarvester approach including the finding that participants considered the system to promote a more flexible and dynamic way of working and increase user enjoyment.

5.4   SUMMARY

Just as people 'design' their information tools, so information tools can be designed for them, based on an understanding of what they need. However, to support a full range of information interactions, we must extend our thinking to consider a range of events, actions and thought processes. In this chapter, we have presented examples of what is possible. Much of this work is at an early stage of development. The important point is that these examples illustrate how the Information Journey can frame thinking about design possibilities that go beyond information provision to support interpretation and use.

CHAPTER 6

Studying User Behaviors and Needs for Information Interaction

In the previous three chapters, we have presented aspects of the context for information interaction and examples of designing for elements of the information journey. In this chapter, we present a view on how to evaluate systems that support information interaction. This shares much in common with approaches to evaluating *any* system, and might be regarded as a personal 'take' on user-centred design and evaluation.

Just as the information journey typically starts from either a need or an opportunity, so the design of a novel system to support information interaction typically starts with either a need, as in most of the cases discussed in Chapter 5, or an opportunity (Blandford and Bainbridge, 2009). There are many descriptions of system development lifecycles, some focusing primarily on stages of system development (e.g., Boehm, B. (1988)) and some more on accommodating user concerns within system development (e.g., Hartson and Hix (1993)). Sharp et al. (2007) present a view of the system development lifecycle as comprising four stages: identifying user needs (i.e., requirements), system design, implementation, and evaluation. These stages are interleaved and iterated to move from early prototypes to final implementation. Carroll and Rosson (1992) discuss the idea that a new system is developed or an existing system adapted, to better satisfy user needs, but that the possibilities offered by a new system often result in new user behaviors, resulting in a co-evolution of system design, adoption and use. Consequently, the activities of identifying user needs and evaluating a system implementation are closely aligned. Indeed, the activity of evaluating an implementation is often a good source of new requirements, as the implemented system can serve as a 'probe' for better understanding the problem.

In the two approaches presented in this chapter, we take the view that there are few differences between conducting studies for requirements gathering and conducting them for evaluating a particular system. In other words, the focus here is on understanding user needs and how a particular system might fit them rather than on the detailed usability of a particular interface. We have adopted this approach based on our own experience of evaluating digital libraries (Blandford et al., 2004); as discussed in Section 2.4, established inspection-based usability evaluation techniques such as Heuristic Evaluation (Nielsen, J., 1994) and Cognitive Walkthrough (Wharton et al., 1994) helped to identify surface-level usability issues but not the deep conceptual difficulties that users experienced when working with digital libraries (e.g., Blandford et al. (2001)).

In this chapter, we present two approaches to conducting user studies for information interactions. Each is intended to address a different set of concerns. The first, which we call PRET A Rapporter, is a generic framework for planning user studies (which might be applied to any system, but it is illustrated with examples of information interactions). This framework is essentially a high-level 'roadmap' that guides planning and executing a user study. Its orientation is practical.

A study may also adopt a conceptual or theoretical framework. A conceptual framework is a set of concepts which guide thinking about the study subject-matter, usually in some theoretically informed way. Such frameworks delineate the object of research and help the researcher converge rapidly on particular phenomena which are important to consider. A conceptual framework defines some important things to study and in doing so offers a cost/benefit payoff on researcher effort. At the same time, it will also constrain scope.

The second approach, which we call Conceptual Structures for Information Interaction (CSII), is such a conceptual framework. It is concerned with gathering and analysing user data in a way that explicitly focuses attention on users' concepts. Where there is an existing system to evaluate, it addresses the quality of the fit between users' conceptual structures and those implemented in the system. Together, PRET A Rapporter and CSII provide practical and theoretical guidance for user studies of information interaction.

6.1   PRET A RAPPORTER: A FRAMEWORK FOR PLANNING A USER STUDY

In order to conduct any study of people working with technology, it is necessary to plan, taking account of practical considerations such as what questions the study is to address and what resources are available. PRET A Rapporter is a framework for planning such studies. This framework is adapted from the DECIDE framework of Sharp et al. (2007), addressing what we perceived as some shortcomings of DECIDE based on our experience (Blandford et al., 2008a)). The stages for designing an evaluation study using the PRET A Rapporter framework are as follows:

1. Purpose of evaluation: what are the goals of the study, or the detailed questions to be answered in the study?

2. Resources and Constraints: what resources are available for conducting the study and, conversely, what constraints must the study work within?

3. Ethics: what ethical considerations need to be addressed?

4. Techniques for gathering data must be identified.

5. Analysis techniques must be selected.

6. Reporting of findings is (usually) the final step.

These stages are not sequential, but generally interleaved and interdependent. We outline each stage in more detail, including brief examples of what is involved in each.

6.1.1   PURPOSE OF EVALUATION

An evaluation should always answer a question. Many decisions such as where and how to gather data and how to analyse data once gathered arise naturally from this initial question. In considering the purpose of an evaluation, an initial consideration is likely to be whether the evaluation is formative or summative – that is, whether the evaluation is to discover new requirements which will inform further design activity, or whether it is to compare one or more systems against an established set of success criteria (e.g., comparing performance measures for alternative system implementations). When identifying user needs, there may not even be a particular system under review; the focus might simply be on understanding user behaviors and needs.

In a research context, a related concern is whether an evaluation focuses on hypothesis testing or developing a deeper understanding of system use in context. For example, McCown et al. (2005), who compared perceptions of the National Science Digital Library (NSDL) with those of Google, may have started with a hypothesis that NSDL supports the science curriculum better than Google. In contrast, Kuhlthau and Tama (2001) (see Section 2.4) were interested in understanding lawyers' tasks, how they use information to accomplish their work, and the role mediators play in their information seeking and use. Consequently, they conducted their study in a largely exploratory way that helped identify requirements for future systems. Where studies are exploratory, aiming to understand a situation better in order to inform the design and deployment of future systems, there will be initial themes to focus data gathering, but data gathering will be open to new possibilities around those themes.

In many situations, it is necessary to refine a theme into a set of questions that will provide a more detailed focus for data gathering. For example, in one study (Stelmaszewska and Blandford, 2002), where the overall goal was to better understand how naïve searchers formulated their queries (with a view to providing better support for this activity), one question was how often each user reformulated their query before adopting a different search strategy or giving up.

In summary, every user study has a purpose, which may be more or less clearly defined. It is helpful to articulate that purpose so as to focus subsequent data gathering and analysis.

6.1.2   RESOURCES AND CONSTRAINTS

Any evaluation study has to work within what is practicable. In commercial contexts, many of the constraints are imposed by contractual considerations, such as the budget available, the timescale within which findings must be reported, and the form of report required. In research settings, these particular resources may be of less immediate concern than others.

One consideration is what system representations are available. For a user needs study, it may be most appropriate to work simply with the running systems to which the users already have access. For evaluations of particular systems, there may be various representations, including running systems or early prototypes; it may also be possible to have access to developers, documentation or source code which facilitates better understanding of (for instance) how IR algorithms have been implemented, or the designers' reasons for providing particular features.

Another central set of questions are where suitable participants can be recruited from, what tasks (if any) they should be given, and what environment they can be studied in. It is usually important to work with participants who represent the intended user population; for example, it would rarely be appropriate to recruit undergraduates in computer science or information studies if the purpose of the study is to evaluate the effectiveness of a specialist medical or law library for supporting practitioners' work.

Another issue is what facilities are available or appropriate for conducting studies and gathering and analysing data. If control over the setting is more important than ecological validity, a study might be run in a usability laboratory, with automatic key-press logging, screen capture and audio and video recording. If higher ecological validity is needed, it may be necessary to visit participants in their work places, which limits what kinds of data gathering are possible.

Others who have reported on digital library evaluations (e.g., Hsieh-Yee, I. (1993); Erdelez, S. (2004)) have given their participants tasks tailored to the purpose of the study, typically including some tasks that have a single, well defined answer and others that require more sophisticated information seeking. Borlund, P. (2003) advocates embedding tasks within a scenario of use. A scenario attempts to artificially re-create an activity context. Implicit in this is that the scenario should specify those aspects of a context which will result in ecologically valid task performance. Whilst assumptions may need to be made about what those factors are, a number of strategies can be used to maximise ecological validity. Attfield et al. (2008b), for example, developed a set of detailed scenarios for use in a lab-based news-writing task based on the findings of a prior field study at The Times (reported in Attfield and Dowell (2003)) and validated by an Executive Editor at the paper.

In some studies, we have favoured less well specified goals in which participants bring tasks which are meaningful to them; for example, when our participants have been postgraduate students, we have often asked them to articulate an information need relevant to their current research project and to search for information to address that need.

A third set of questions relate to expertise and support for data analysis. Often, the central question is what expertise is available in applying different techniques. For example, it would be unwise to plan a sophisticated quantitative study if there were insufficient expertise in experimental design and statistical analysis available. If new techniques need to be learnt in order to conduct an evaluation, it is necessary to consider what resources there are to learn those techniques.

In particular settings, there may be other constraints to consider; for example, in a study of use of DLs by clinicians (Adams et al., 2005b), timing and location were important: one interview with a surgeon was held in the surgery ante-room between operations, and focus groups for nurses and allied health professionals (e.g., nutritionists, physiotherapists) were held at the end of meetings because they found it very difficult to find time to participate in the research individually.

6.1.3   ETHICS

In evaluation studies, it is important to consider ethical dimensions. Most professional bodies (e.g., ACM (1999)) publish codes of practice. Less formally, we have identified three important elements of ethical consideration:

- **V**ulnerable participants (young, old, etc.).

- **I**nformed consent.

- **P**rivacy, confidentiality and maintaining trust.

Since many studies of information interaction involve people at work, they do not involve participants that might be regarded as vulnerable; counter-examples include the work of Theng et al. (2001); Druin, A. (2005) and Bilal and Bachir (2007) on designing digital libraries with and for children, Livingstone and Bober (2004) on children's use of the internet, and Aula, A. (2005) on the information seeking of older users. Our studies of digital library use in clinical settings (Adams and Blandford, 2002; Adams et al., 2005b) included observations of clinical encounters between doctors and patients, for whom privacy and confidentiality were paramount concerns. In this case, anonymisation of data, for both individuals and institutions as a whole, was imperative – both for protecting privacy and also to maintain individuals' and organisations' trust in the research procedure.

It is now recognised as good practice to ensure all participants in any study are informed of the purpose of the study (e.g., that it is the system that is being assessed, or their work that is being understood, and not them) and of what will be done with the data. Also, participation should be voluntary, with no sense of coercion (e.g., by the exercise of a power relationship between evaluator and participants).

Data should normally be made as anonymous as possible, and individuals' privacy and confidentiality need to be respected. While immediate respect of individuals is reasonably obvious, less obvious is the need to continue to respect participants' privacy in future presentations of the work and to show similar respect to groups and organisations. Lipson, J. (1997) discusses many of the less obvious pitfalls of publishing the findings of studies, such as participants feeling betrayed or embarrassed by descriptions of their behavior or attitudes even if those descriptions are anonymised.

Mackay, W. (1995) discusses 'best ethical practice' for researchers with personal participant data. She notes that professional ethics should ensure that multimedia data is used within acceptable boundaries. She also proposes that individuals' identities should be hidden, wherever possible, during recording. Adams, A. (1999) highlights the importance of participants' awareness of who is seeing their information, in what context, and how they will use it.

6.1.4   TECHNIQUES FOR DATA CAPTURE

Ethical considerations cover all aspects of a study, including data collection, analysis and reporting. While these steps may be interleaved (particularly in large studies), we consider them in order.

Techniques for data collection cannot be addressed completely independently of practical constraints or intended analysis techniques; nevertheless, the purpose of the evaluation will inform what data collection techniques are likely to be appropriate.

Evaluation questions that involve counts of events or a test relating independent and dependent variables will clearly demand that appropriate quantitative data be gathered. Within studies that focus more on user behavior, numerical data may include numbers of particular event types or user action types, or times to perform tasks. Such data is most commonly captured using some form of computer logging (e.g., Nicholas et al. (2006)).

In evaluating users' experiences of working with information, we have focused much more on qualitative data. Nevertheless, we have recruited a variety of data collection techniques including naturalistic observations, think-aloud protocols, in-depth interviews, access to server logs, and focus groups, as appropriate to the particular questions being addressed in each study. For detailed descriptions of such qualitative data collection approaches, see texts such as that by Kuniavsky, M. (2003).

6.1.5   ANALYSING DATA

Quantitative data is typically analysed using statistical techniques (or simpler reports of numbers). For quantitative usability evaluation, standard psychology statistics texts such as Pagano, R. (2001) are a useful resource.

Qualitative data analysis may take many forms, as described by Miles and Huberman (1994). Depending on the research question, the analysis may vary between more focused issues (e.g., what errors do users make with this system and how might they be avoided?) to exploratory concerns (e.g., how do users work with information in this work setting and how might their work be improved through system design?). The main data analysis technique we have employed in our studies is Grounded Theory (Glaser and Strauss, 1967). This is a social-science approach to theory building that can incorporate both qualitative (e.g., interviews, focus groups, observations, ethnographic studies) and quantitative (e.g., questionnaires, logs, experimental) data sets. The methodology is inductive, being driven by the data and a desire to discover, rather than test any particular a priori hypothesis. It combines systematic levels of abstraction into a framework about a phenomenon, which is iteratively verified and expanded throughout the study. We have also adapted the approach to relate it to relevant theoretical perspectives – e.g., Adams et al. (2005b) related findings from a study to established theory on Communities of Practice (Wenger, E., 1999), as discussed in Section 3.2.2.

6.1.6   REPORTING FINDINGS

The final step is reporting findings. In research projects, the main means of reporting is normally through academic publications presenting the aims (or purpose) of a study, the background, the method or methods applied, the results, and conclusions (often including a discussion of implications for design). For quantitative experiments, it is customary to provide information at a level of detail that would enable others to replicate the study to establish its reliability. For qualitative studies, it is rarely possible to replicate the conditions of a study closely enough to expect this degree of reliability, so the focus is usually more on presenting the method, analysis and findings in sufficient detail to enable the reader to assess their validity.

In interacting with developers and policy makers, less formal reporting channels are usually appropriate. These might include executive summaries that focus on problems found and possible design solutions proposed.

6.1.7   SUMMARY

In summary, the PRET A Rapporter framework provides a checklist for planning an evaluation study. It is a general framework and practical guide that can be applied to any interactive system user study. The second technique we discuss, CSII, involves the application of a conceptual framework, which focuses the analyst's attention on the quality of the fit between users' conceptual structures and those implemented in the system.

6.2   CONCEPTUAL STRUCTURES FOR INFORMATION INTERACTION (CSII)

For understanding how people interact with information, there are many possible concerns: the conceptual structures people work with, the processes they follow, the graphical structures they work with, how information is communicated, etc. In this section, we focus particularly on conceptual structures, since these have received relatively little attention in the past, but are important for information interaction. To understand user needs, the primary concern might simply be with people's conceptual structures of a particular information domain and considering the implications for design. When evaluating a particular system, one can go further and assess the quality of the conceptual fit between user and system.

To take a simple example, in a study of ambulance control, Blandford et al. (2002) found that controllers plan their work around "calls." However, a deeper analysis of these "calls" revealed that two concepts were being merged: sometimes, they really were talking about a "call," a particular interaction with a caller about an emergency; more often, however, they were actually talking about an "incident," an event to which an emergency response was needed. Historically, there would have been a one-to-one mapping between calls and incidents, but with the rise in the use of mobile phones, it is now very common for them to receive multiple emergency calls about the same incident if it takes place in a public place. This highlights the need for controllers to have a system that allows them to manage incidents and relate each incident to all call information about it. The computer aided despatch system in use at the time of our study supported the management of calls, rather than incidents; when there were multiple calls about an incident, the system did not support controllers in relating the (often incomplete or imprecise) information from different calls about the same incident, increasing their workload and making the overall system more prone to error.

To support analysis of this kind of misfit between the ways users and systems conceptualise information structures, the CSII (Conceptual Structures for Information Interaction) approach has been developed. This is a simplification of the CASSM approach (Blandford et al., 2008b). CSII supports designers and evaluators in assessing existing information interaction designs and in developing new designs that better fit users' needs. Evaluation of existing systems directly informs re-design. It does not necessarily dictate or even indicate, the precise form of that re-design, but it highlights the possibility in the form of user-centred requirements.

CSII focuses attention on the conceptual structures that people are working with. If a particular system is being evaluated, the representations implemented in the system will be compared against these. Sometimes, these are unavoidably different; for example, when probing researchers' conceptualisations of the research literature (Faisal et al., 2006), we found that important concepts for people were the "idea" that a paper represents, the "area" that another researcher works in, and the "community" of researchers working on a particular topic. These concepts are not readily implemented in a system, although new social networking tools may make it possible to represent these concepts more explicitly in the future. Sometimes, as in the ambulance control example outlined above, it would be fairly straightforward to extend the system to directly represent information about other key user concepts, such as incidents.

The process of conducting a CSII analysis involves first gathering user-centred data and identifying requirements from it. If an existing system is being evaluated, it is also necessary to gather corresponding system-centred data and compare the two to assess conceptual fit.

6.2.1   CSII FOCUSES ON CONCEPTS

The focus for conducting a CSII analysis is on *concepts*.

A *concept* is a "thing" or a "property" that the user works with while interacting with information. These can be divided into *entities* and *attributes*, although early on in analysis, this distinction may not be important.

An *entity* is often something that can be created or deleted within the system. Sometimes, entities are things that are there all the time, but that have attributes that can be changed. In the ambulance dispatch example discussed above, entities include incidents and calls.

An *attribute* is a property of an entity – usually one that can be set when the entity is initially created or that can be subsequently changed. For example, an incident has attributes including the location and nature of the incident, and whether or not it has been dealt with.

CSII for requirements focuses on identifying user concepts and the ways in which those concepts are manipulated through interacting with information. There is a need to consider which can be implemented in a system and how users might need to manipulate or relate information objects. Here we focus more on the use for evaluation of an existing system.

If evaluating an existing system, there may be system concepts that users have to be aware of and work with, that are not an immediate part of the users' conceptualisation. In this case, for every concept, the analyst determines whether it is *present*, *difficult* or *absent* for the *user* and in the *system*. Concepts that are present in one place but absent from the other are sources of potential misfits, as are concepts that are *difficult*.

The stages of a CSII analysis are first gathering data; then identifying user concepts. If CSII is being used for evaluating an existing system, it is also necessary to identify the corresponding system concepts, identify misfits, and analyse the actions as implemented in the system. Here, we present each of these stages, followed by an example analysis taken from a study of digital libraries.

6.2.2 GATHERING USER DATA

For gathering user concepts, some form of verbal data is required from users. This might be from a think-aloud protocol (of the user working with a current system), from Contextual Inquiry interviews (Beyer and Holtzblatt, 1998), from other kinds of interviews, or from user-oriented documentation – for example, describing formalised user procedures for completing tasks. In general, more sources of data yield more information, but this has to be balanced against any need for speed, efficiency or the practicalities of accessing different sources. Data sources should give information about how the system is used in its normal context of use. Relating back to the PRET A Rapporter framework, the *purpose* of a CSII study is to gather users' conceptualisations of the domain they are working with; for this, appropriate data gathering techniques are highly contextualised to the user's domain of work, and analysis has to be qualitative, focusing on users' conceptual structures.

For a CSII analysis, it is particularly important that the user data that is gathered gets to the heart of how the user thinks about their domain activities and not just how they might interact with a particular device.

If interviews are being used to gather user data, then a semi-structured interview format generally works most effectively. This involves preplanning a broad interview agenda, to establish what people are achieving (or hoping to achieve) when working with the (current or future) system. Questions need to be designed to probe the concepts that people are working with, how the system changes the state of the world, or the user's knowledge. Ideally, the interviewer will have key questions pre-planned but respond to the individual in the context of the interview to probe interesting avenues more deeply.

If think-aloud data is being gathered to support analysis, based on how people use a current system, then it is important that the tasks given to study participants are domain-relevant and give the participants scope for discussing domain concepts. For example, the analyst evaluating a shopping website would get little useful data if the user task were given as "use the XX grocery site to buy three pints of milk and a loaf of bread using the search facility." A task description such as the following would yield more useful data: "You have invited some friends round for supper, and are hoping to be able to order everything you need from the XX online store. Talk through how you would do this, using the site to put all the items you require in a basket. Stop before you get to the payment part of the process." This version of the task description might allow people to show how they plan the catering for a small event, whether they refer to recipes to support their planning, how they think about the organisation of the shopping, etc.; these are the kinds of issue that matter for a CSII analysis.

Contextual Inquiry interviews (Beyer and Holtzblatt, 1998) are an effective way to mix interviews and observations in data gathering, particularly, if the system under investigation is one that is used in a work context. This allows participants to articulate both what they are doing (in broad terms) and how they are using a particular system to achieve it. The observer's role is to ask probing questions that elicit the participant's understanding of the broader activity of which their interaction with a particular system is a part.

The most effective data gathering often involves using multiple methods. For example, interviews can be used to find out generally what people do in, and think about, a particular activity, which can be used to design suitable tasks for a think-aloud study that elicits more information about how people perceive a particular system for achieving those tasks (assuming that an implementation already exists).

6.2.3   IDENTIFYING USER CONCEPTS

The next step is to identify core user concepts from the data. One way to conduct an analysis is to go through the words (e.g., transcription of users talking or documentation) highlighting nouns and adjectives, then deciding which of those words represent core concepts within the user's conceptualisation of the domain (and, if appropriate, the system they are working with).

Depending on what matters most, the analyst might distinguish between entities and attributes to achieve clarity in the model. Concepts might also be grouped into related ones that function together or that might be displayed together.

The process of identifying concepts is exemplified in the case study presented below (Section 6.2.7).

6.2.4   IDENTIFYING SYSTEM CONCEPTS FOR EVALUATING AN EXISTING SYSTEM

If the analysis is an evaluation of an existing system, access to a system description is also needed. The main sources of system concepts are system descriptions and maybe a running system. Again, the data is analysed in whatever ways are possible (depending on what data sources are available) to identify core system concepts.

In doing this analysis, one thing to avoid is extensive descriptions of interface widgets; rather, the analysis should focus on the underlying system representation. Interface widgets are a means to an end, not an end in themselves. For example, an analyst describing a telephone would focus on calls, people, numbers and the line state, not on the keys on the keypad or the receiver as an object.

6.2.5   IDENTIFYING MISFITS

Once suitable data has been gathered, misfits can be identified. The first step is simply to identify and compare system and user concepts; a second stage of analysis considers what actions are needed to change the system, and whether there are problems with actions (see below, Section 6.2.6).

Misfits between user and system are probably the most important information-related misfits. These misfits fit into three classes:

User concepts that are not represented within the system, and hence they cannot be directly manipulated by the user. A very simple example is the use of land-line telephones: the user is normally interested in speaking to a particular individual, but the only means of doing that it to place a call to a location where they are likely to be (e.g., their home or office) because the telephone system does not map handsets to individuals. Mobile telephones, that are generally assigned to one individual, and for which the user only has to make the simpler mapping of name to number, partly overcome this misfit, though they can suffer from a converse difficulty which is the user distinguishing between 'work calls' and 'domestic calls', and treating these differently – whether in the way incoming calls are responded to or outgoing calls are billed. For some users, this misfit is sufficiently important that they develop a workaround of carrying two handsets with them to distinguish between different types of calls.

User concepts that are not represented in the system often force users to introduce workarounds, as users are unable to express exactly what they need to, and therefore use the system in a way it wasn't designed for. Other examples include using a field in an electronic form to code information for which that form was not actually designed or keeping paper notes alongside an electronic system to capture information that the system does not accept.

System concepts that the user has to know about but that are not naturally part of their initial understanding and, therefore, need to be learned. An example involving information structures might be the ways that information is organised in a hierarchical classification system.

For users, these misfits may involve simply learning a new concept, or they may involve the users constantly tracking the state of something that has little significance to them.

User- and system concepts that are similar but non-identical and which are often referred to by the same terms. This could be considered as an amalgamation of the two categories above (a user concept that the system doesn't represent and a system concept that the user has to know about), but has a particular set of implications, in terms of how the user has to mould their understanding to the system. One example is the difference between a call and an incident in ambulance dispatch, as discussed above.

These misfits may cause difficulties because the user has to constantly map his / her natural understanding of the concept onto the one represented within the system, which may have a subtly different set of attributes that the user then has to work with.

As well as concepts being absent, some may be available but present some kind of difficulty. For users 'difficult' concepts are most commonly ones that are *implicit* – ideas they are aware of if asked but not ones they expect to work with. An example, for many people, is the end time of a meeting in a diary system: in people's paper diaries, many engagements have start times (though these are often flagged as 'approximate' – e.g., '2ish') but few have end times, whereas electronic diaries (which are sold as diaries, but are better described as scheduling systems) force every event to have an end time (or a duration, depending on how you look at it). This forces users to make explicit information that they might not choose to. Of course, there are (typically busy) people for whom the "scheduling" nature of electronic diaries suits them better than the relatively imprecise structure of paper diaries (Blandford and Green, 2001), but these are a minority of users.

Other sources of difficulty might be that the user *has to learn* the concept, or that it is perceived as *irrelevant* by the user. A system concept may be difficult to work with by being *disguised* (represented at the interface, but hard to interpret); *delayed* (not available to the user until some time later in the interaction); *hidden* (the user has to perform an explicit action to reveal the state of the entity or attribute); or *undiscoverable* (findable by the user who has good system knowledge, but unlikely to be discovered by most users). Which of these apply in any particular case – i.e., why the concept might cause user difficulties – is a further level of detail that can be annotated by the analyst.

6.2.6   ADDING IN INFORMATION ABOUT ACTIONS

Because CSII is primarily concerned with conceptual misfits, actions are of secondary concern. The analyst can define how actions change the existence of entities or the values of attributes as a further step of analysis.

An action that is difficult to perform for some reason is referred to as being *hard*; maybe it involves a long and tedious action sequence or it is difficult for the user to discover. An action that is impossible, but that the analyst thinks the user might want to do under some circumstances, is referred to as *can't*. (This contrasts with *fixed*, which is an action that is impossible, but not judged to be problematic.) The other possibilities are that the action is *easy*, or that it is done by the *system* (which may include other agents – e.g., over a network, or simply other people); again, many of these cases are not actually problems, and it is up to the analyst to consider implications.

As well as describing actions in the above terms, the analyst should also be alert to changes that are impossible, side effects of actions, and actions that have unpredictable effects, and note any misfits identified while they are working through the analysis.

6.2.7   WORKED EXAMPLE: A DIGITAL LIBRARY SYSTEM

To illustrate the approach, we present brief examples from an analysis of how people work with a digital library. This example is presented in more detail by Blandford et al. (2008a).

Data was gathered from four students completing a Masters course in Human–Computer Interaction and all working on their individual research projects at the time of the study. These participants are representative of one important class of users who make regular use of digital libraries. Participants were invited to search for articles relevant to their current information needs and to think aloud while doing so. Occasionally, the observer intervened with questions in the style of

Contextual Inquiry (Beyer and Holtzblatt, 1998) to encourage participants to talk more explicitly about their understanding of the systems they were working with. Participants were invited to work with whatever information resources they chose; in practice, all four chose to work mainly with the ACM DL, which provides an opportunity to include a brief evaluation of this digital library in our discussion. Audio data was recorded and transcribed, and notes were made of key user actions.

In this analysis, five categories of user concepts were identified:

- Concepts concerning the information they were looking for, relating to the topic of the search, the domain within which they were searching, particular ideas they were interested in and particular researchers who were pertinent to their interests.

- Features of the resources (the ACM DL, other digital libraries, the HCI Bibliography, Google and the Web of Knowledge) that they used.

- Concepts from the domain of publishing, relating to journals, articles, publishers, authors, etc.

- Concepts pertaining to electronic search, such as query terms, results lists and their properties.

- Features of themselves as consumers of information, including their interests, domain knowledge, search expertise, access rights to particular resources and research direction.

To illustrate how verbal protocol data feeds into analysis and helps identify new design possibilities, we focus on the first of these categories: how users described what they were looking for.

Participant 1 was looking for fairly focused material:

"I was looking at something on 'language patterns' yesterday on Google, so I'd like to have a look a bit more on that. It was originally thought up by Christopher Alexander, who is usually in design and architecture but I think there might have been studies applying the work to HCI."

He regarded this as "an unspecified topic."

This participant is talking not just about a topic (how pattern languages can be used in HCI), but also about a particular idea (pattern language), the researcher who originated that idea (Christopher Alexander) and domains of research (design, architecture, HCI). He reinforced some of this understanding, saying:

"Christopher Alexander is the guy that has been attributed to coming up with the idea."

And "I'd go to the ACM because it's an HCI topic."

He illustrated his understanding of the cross-referencing of material between texts (that the work of one researcher would be referred to in a work by another), saying that he "came across it by accident in a book by Heath and Luff."

He also recognized that there were different ways of expressing this topic as a search query. For example:

"I know the topic I want is either 'language patterns' or 'pattern languages'."

Participant two also started with not just a topic (focus groups) but the view that the material of interest would pertain to a particular domain (HCI):

"I was thinking of looking for how focus groups are used in HCI in terms of evaluating initial designs and stuff like that."

She recognized that she was uncertain about how to describe this topic to the search engine:

"I'd put in "focus groups" using quotes so that, well I'm guessing that it deals with it as one word and, I'm not sure. I'll put a plus. I'm never quite sure how the search, I wouldn't say random, but you seem to get funny stuff back from the ACM in terms of results."

She also recognized that there was material returned in response to a search that was of more or less interest to her:

"I'd just go through some of this stuff and see what I find interesting. Year 2000 focus groups. I'm thinking that's about the year 2000 bug, so it won't be that relevant."

Participant 3 was also concerned with the interestingness of particular topics:

"I'm specifically interested in designing interactive voice systems, so I wouldn't be interested in a lot of these titles."

He had some difficulty identifying suitable search terms and, like participants 1 and 2, was concerned about finding articles in the domain of HCI:

"I'll change it to 'voice response usability' because 'usability' is fairly specific compared to 'design', which could be technology, or architecture: it could be in a different domain."

Participant 4 expressed an understanding of broader and more narrow topics as a focus for search:

"I'm going to look for something on qualitative data analysis as a broad thing and see if I can get something to do with dialogue analysis or something to do with conversation."

Participant 4 echoed themes from earlier participants concerning topics and interestingness:

"This one says something about dialogue structure and coding schemes and although it's not really to do with my topic, it might be useful."

She also indicated that she was exploring how to describe the interesting topic in search terms:

"I'm going to go back to my original search list and put in 'dialogue' and 'coding' because I hadn't thought about looking for that term."

These extracts show how the same concepts are expressed (albeit in slightly different ways) by multiple participants, giving assurance that they are significant concepts from a user perspective. In summary, the key concepts identified were:

- Topic, with attributes:

  - Specificity.

  - What domains it features in.

  - How expressed.

  - Interestingness.

- Idea, with attribute:

  - Originator.

- Domain.

- Researcher, with attributes:

  - Name.

  - Works in domain.

  - Generated idea.

  - Wrote paper.

  - Referenced in (other) paper.

These concepts can contribute to system design by being represented in a notation chosen for system development (a topic that is beyond the scope of this lecture). They can also be the starting point for evaluating an existing implementation, such as the version of the ACM digital library that was available at the time of this study.

To form the assessments relating to the system, it was necessary to look at the ACM DL and its interface. Findings relative to the system are presented in Table 6.1.

Here, we see that many central user concepts are absent from the system representation. For example, there was no direct representation of topics, specific ideas or domains in the underlying system. It may be that by reading the interface representation, such as the titles or abstracts of papers, the user could infer the topic, ideas and domain, but they are not clearly represented. The user is forced to find work-arounds for expressing these concepts. If we consider how they might be represented in the DL (e.g., exploiting existing meta-data), we realise that these ideas are similar

Table 6.1: Concepts Relating to the Topic of a Search		
Entity / attribute	**User**	**System**
Topic	present	absent
Specificity	present	absent
Features in domains	present	absent
How expressed	present	absent
Interestingness	present	absent
Idea	present	absent
Originator	present	absent
Domain	present	absent
Researcher	present	present
Name	present	present
Works in domain	present	absent
Generated idea	present	absent
Wrote paper	present	present
Ref'd in (other) paper	present	present
Collaborated with	absent	present

(but not identical) to the 'general terms' and 'index terms' in the ACM classification system. We could imagine an implementation of the search facility that allowed the user to easily prioritise (for example) articles that included the general term "Human Factors." The current implementation of the ACM DL hides this possibility, and none of our participants discovered it. Participant 3 expressed this succinctly:

> "ACM's big, it's got all the different disciplines within it, so I'm just trying to focus in on usability related stuff."

This indicates a promising design change: to make it easier for users to focus a search by particular general terms or classifiers within the ACM classification system (or, indeed, to be able to browse by classifier).

One concept that the DL represents comparatively well is that of a researcher (or at least, an author). Indeed, although information about what domain an individual works in and what ideas they have originated is poorly represented, other information about them – notably, people they have co-authored with – is clearly represented through a link to "collaborative colleagues." Although none of the participants in this study referred to the idea that an individual might collaborate with others, or hinted that might be an interesting or important item of information, this information may be of use to other user groups of the ACM DL (this, of course, is an empirical question). In this case, the representation of a concept (who collaborates with whom) within the DL does not represent a difficulty to the user because the use of this feature is discretionary.

In this example, we have shown how user data can be used to identify requirements on an interactive system and how that data can be compared against an existing design or implementation to evaluate that design.

6.2.8   DIFFERENT USER GROUPS

In this case, we created a 'user' description by merging the descriptions from several individual users who have similar perceptions of the system with which they work. A system very often has different stakeholder groups who have different information needs which need to be analysed separately. Sometimes, the different groups necessarily need to use the same system but possibly via different interfaces (e.g., Cooper, A. (1999) discusses interfaces to an in-flight entertainment system for use by passengers and crew). Sometimes, there are stakeholders who are not necessarily system users but who have a strong influence over system design; for example, advertisers may wish to tell people things that they didn't necessarily want to know about during the interaction, and the challenge of good design is to make advertising acceptable and effective (i.e., to persuade people to expand the range of things they are interested in).

Sometimes, there are simply different user groups, who use a generalised system that could, if desired, be redesigned as several more specialised systems that might be easier to use for a more narrow range of tasks.

6.2.9   SUMMARY OF CSII

In summary, application of the CSII conceptual framework can help identify user needs related to information structures and the concepts that users are working with. This can be used to inform the conceptual design of a new system or the evaluation of an existing system in terms of conceptual misfits. The primary concern of the approach is with entities and attributes and their properties. Actions are considered as part of evaluation if required. CSII is intentionally sketchy. There is no unique 'right answer' in a CSII analysis; rather, the aim is to simply focus the analyst's attention on a set of things that matter most to users, and for design.

6.3   SUMMARY

There are many excellent texts on evaluating systems in general. In this chapter, we have focused on approaches to the challenge of understanding people's interactions with information and evaluating systems to support those interactions. The first was a general checklist for planning an evaluation study; the second, CSII, was a conceptual framework oriented towards analysing users' conceptual structures, for both design and evaluation.

CHAPTER 7

Looking to the Future

Most effort on supporting information working to date has focused either on information generation (e.g., word processing) or information provision. There is an apparently widely held view that information provision is sufficient for supporting information work. This view contrasts with the lived experience; the effective provision of information is necessary but not sufficient to make that information useful and usable.

Classical approaches to Information Retrieval typically support well defined information needs well. They do not, however, support the much richer information behaviors that people regularly exhibit as they go about their work and leisure activities. There is a need to distinguish between routine and sophisticated information activities, and to develop a richer repertoire of ways of supporting the latter.

We have illustrated possibilities that are currently at the prototype stage of development. These prototypes support key stages of the information journey, namely sensemaking and information interaction and use. These more sophisticated design solutions are not going to be "one size fits all," but will need to recognise and respond to the much richer repertoire of information behaviors that people engage with in different contexts.

Looking to the future, where are the big gains? It does not seem to be in newer technologies necessarily superseding established ones. For example, while the quill pen has been largely superseded by newer technologies, the pencil still has a valuable role to play (Duguid, P., 1996). The trends appear to be towards more mobile and distributed information interactions and to new ways to manage the volume of information that is available. New technologies such as e-books and large displays may augment the range of information interactions that are possible, but are unlikely to supersede existing technologies in the foreseeable future. Information interaction does not take place in isolation, but it is a situated, and usually social, activity.

As should be clear from the work presented in this lecture, active information finding will remain essential to many information interactions; the act of finding offers too many opportunities for comprehending the information landscape and reflecting on the understanding of the information problem to be delegated to information professionals or systems.

One of the recognised challenges of finding information is that simple string-matching (searching for specified query terms) is limited: many words have multiple meanings, resulting in poor precision, while many ideas (or information needs) can be expressed in alternative ways, resulting in poor recall. As an attempt to address these limitations, there is ongoing research on the "semantic web," for which a focus is on making information "machine comprehensible," and thereby improving the quality of information returned from searches.

In addition, one can imagine some relatively well-defined information needs that can be comprehended and addressed automatically by systems – e.g., through a subscription service. One approach may involve more automated 'background' information finding by systems to present information to people with less effort on their part – whether in response to long-term information interests (e.g., Farooq et al. (2008); Adams et al. (2005b)), to immediate information needs (e.g., Twidale et al. (2008)) or serendipitously addressing latent information interests (e.g., Toms and McCay-Peet (2009); Foster and Ford (2003)).

There is also a clear need for richer approaches to supporting information interpretation. While it will become increasingly common for routine information interpretation to be available automatically, this will need to become increasingly sophisticated and tailored to the needs of the individual, e.g., based on education, prior knowledge and reasons for wanting the information. There will always be a need for experts who are creating new understanding and interpreting that understanding for use by others.

We have highlighted the opportunities presented by better integration between information seeking and use. To date, most of the research in this area has focused on writing as the principal use of information, but we anticipate that other uses (design, planning, decision making) may become more explicitly supported in future.

In this lecture, we have focused mainly on productive work settings rather than the role of information interaction in leisure activities. This is partly because needs are less well articulated and, therefore, harder to study in leisure settings. There is a need for more research on information behaviors in leisure contexts and for novel systems to support those behaviors as well as work-related behaviors.

At present, resources are being wasted on inappropriate technologies, based on incorrect assumptions about information practices. A richer understanding of what people really do and why, should inform the design of innovative technologies to support those activities.

CHAPTER 8

Further Reading

Research on interacting with information appears in many different journals and conferences, depending on the research tradition of the authors. Conferences where relevant work might be published include the ACM / IEEE Joint Conference on Digital Libraries (JCDL), the European Conference on Digital Libraries (ECDL), Information Seeing in Context (ISIC) and ACM Computer–Human Interaction (CHI). Journals include the Journal of the American Society for Information Science and Technology, Information Processing and Management, the Journal of Documentation and the Human–Computer Interaction Journal. Papers may also be found in journals addressing particular domains of application, such as the Law Library Journal or the Health Information and Libraries Journal.

Brown and Duguid (2000) present a rich account of the broader social and organisational context within which information interactions take place, and they highlight many aspects of information exchange and interpretation that can often do unnoticed by managers and technology developers alike.

Hearst, M. (2009) reviews information seeking models in more detail than has been done in this lecture, and he discusses the design and evaluation of search interfaces from a user-centred perspective.

Other Synthesis Lectures on both Human–Computer Informatics and Information Concepts, Retrieval and Services may also be of relevance; in particular, White and Roth (2009) present work on exploratory search, which is an essential component of information interaction and is often overlooked in more traditional approaches to information provision.

As discussed in Chapter 2, information interaction spans traditional disciplines, and it has much to learn from and contribute to all those disciplines.

Bibliography

ACM (1999). *Software Engineering Code of Ethics and Professional Practice.* http://www.acm.org/serving/se/code.htm (accessed 31/8/09) 6.1.3

Adams, A. (1999). Users' perceptions of privacy in multimedia communication. *Proceedings of CHI' 99*, Pittsburgh. ACM Press. 53–54. DOI: 10.1145/632716.632752 6.1.3

Adams, A. and Blandford, A. (2002). Acceptability of medical digital libraries. *Health informatics Journal.* 8(2). 58–66. DOI: 10.1177/146045820200800202 3.2.2, 3.2.3, 3.3.1, 6.1.3

Adams, A. and Blandford, A. (2005). Digital libraries' support for the user's 'information journey.' In *Proceedings of ACM/IEEE Joint Conference of Digital Libraries (JCDL'05).* 160–169. Denver, Colorado. ACM. DOI: 10.1145/1065385.1065424 4.1

Adams, A., Blandford, A., Budd, D. and Bailey, N. (2005a). Organisational Communication And Awareness: A Novel Solution. *Health Informatics Journal*, 11: 163–178. DOI: 10.1177/1460458205052357 5.1.2

Adams, A., Blandford, A., and Lunt, P. (2005b). Social empowerment and exclusion: a case study on digital libraries. *ACM Transactions on Computer–Human Interaction.* ACM Press. 12(2). 174–200. DOI: 10.1145/1067860.1067863 3.2.1, 3.2.2, 3.3.1, 6.1.2, 6.1.3, 6.1.5, 7

Attfield, S. (2005). *Information seeking, gathering and review: Journalism as a case study for the design of search and authoring systems*, PhD Thesis, University of London. 4.2.2

Attfield, S., Adams, A., and Blandford, A. (2006). Patient information needs: pre and post consultation. *Health Informatics Journal.* 12. 165–177. DOI: 10.1177/1460458206063811 3.2.3

Attfield, S. and Blandford, A. (in press) Social and interactional practices for disseminating current awareness information in an organization setting. *Information Processing & Management.* DOI: 10.1016/j.imp.2009.10.003 3.2.1, 4.4

Attfield, S., Blandford A., Dowell J., and Cairns P. (2008a). Uncertainty-tolerant design: Evaluating task performance and drag-and-link information gathering for a news writing task. *International Journal of Human-Computer Studies.* 66. 410–424. DOI: 10.1016/j.ijhcs.2007.12.001 4.1, 5.3.2, 5.3.2

Attfield, S., Blandford, A., and Dowell, J. (2003). Information seeking in the context of writing: a design psychology interpretation of the 'problematic situation.' *Journal of Documentation.* 59(4). 430–453. DOI: 10.1108/00220410310485712 4.6

Attfield, S.J. and Dowell, J. (2003). Information seeking and use by newspaper journalists. *Journal of Documentation*, 59(2), 187–204. DOI: 10.1108/00220410310463860 2.5, 3.2.2, 3.3.1, 4.2.2, 6.1.2

Attfield, S., Fegan S., and Blandford A., (2008b). Idea Generation and Material Consolidation: Tool Use and Intermediate Artefacts in Journalistic Writing. *Cognition, Technology and Work*. Online first February 2008, DOI: 10.1007/s10111-008-0111-6 2.5, 3.3.1, 6.1.2

Attfield, S., Makri, S., Kalbach, J., Blandford, A., DeGabrielle, S., Edwards, M. (2008c). Enquiry prioritisation, information resources and search terms: Three kinds of decision at the virtual reference desk. *Proc. ECDL 2008*. LNCS 5173. 106–116. 3.2.1

Aula, A. (2005). User study on older adults' use of the Web and search engines. *Universal Access in the Information Society*, 4. 67–81. DOI: 10.1007/s10209-004-0097-7 6.1.3

Bates, M. (1989). The Design of Browsing and Berrypicking techniques for the Online Search Interface. *Online Review*, 13,5. 407–424 DOI: 10.1108/eb024320 2.2, 3.3, 4.5

Belkin, N., Oddy, R., and Brooks, H. (1982a). ASK for information retrieval: part I. *Journal of Documentation*. 33(2), 61–71. 2.2, 4.2

Belkin, N., Oddy, R., and Brooks, H. (1982b). ASK for information retrieval: part II. Results of a design study. *Journal of Documentation*, 38(3), 145–164. DOI: 10.1108/eb026726 2.2

Beyer, H., Holtzblatt, K. (1998). *Contextual Design*. San Francisco: Morgan Kaufmann. DOI: 10.1145/291224.291229 4.1, 6.2.2, 6.2.7

Bilal, D. and Bachir, I. (2007). Children's interaction with cross-cultural and multilingual digital libraries: I. Understanding interface design representations, *Information Processing & Management*. 43(1). 47–64. DOI: 10.1016/j.ipm.2006.05.007 6.1.3

Blandford, A., Adams, A., Attfield, S., Buchanan, G., Gow, J., Makri, S., Rimmer, J., and Warwick, C. (2008). PRET A Rapporter: evaluating Digital Libraries alone and in context. *Information Processing & Management*. 44. 4–21. DOI: 10.1016/j.ipm.2007.01.021 2.4, 6.1, 6.2.7

Blandford, A. and Bainbridge, D. (2009). The pushmepullyou of design and evaluation. In G. Tsakonas and C. Papatheodorou (eds.) *Evaluation of Digital Libraries*. Chandos Publishing. 3.3, 6

Blandford, A. E. and Green, T. R. G. (2001). Group and individual time management tools: what you get is not what you need. *Personal and Ubiquitous Computing*. 5(4). 213–230. DOI: 10.1007/PL00000020 6.2.5

Blandford A., Green T.R.G, Furniss D. and Makri S. (2008). Evaluating system utility and conceptual fit using CASSM. *International Journal of Human–Computer Studies*. 66. 393–409. DOI: 10.1016/j.ijhcs.2007.11.005 6.2

Blandford, A., Keith, S., Connell, I., and Edwards, H. (2004). Analytical usability evaluation for Digital Libraries: a case study. In *Proc. ACM/IEEE Joint Conference on Digital Libraries*. 27–36. DOI: 10.1145/996350.996360 2.4, 6

Blandford, A. E., Stelmaszewska, H., and Bryan-Kinns, N. (2001). Use of multiple digital libraries: a case study. In *Proc. JCDL 2001*. ACM Press, 179–188. DOI: 10.1145/379437.379479 2.4, 3.3.2, 6

Blandford, A. E., Wong, B. L. W., Connell, I. W., and Green, T. R. G. (2002). Multiple viewpoints on computer supported team work: a case study on ambulance dispatch. In X. Faulkner, J. Finlay and F. D. Étienne (eds), *Proc. HCI 2002 (People and Computers XVI)*, 139–156. Springer. 6.2

Boehm, B. (1988). A Spiral Model of Software Development and Enhancement, *Computer*, May 1988, 61–72. DOI: 10.1109/2.59 6

Borgman, C. (2003). Designing digital libraries for usability. In A. Bishop, N Van House *et al.* (Eds.), *Digital library use*, MIT Press, Cambridge. 3.1.1

Borlund, P. (2003) The IIR Evaluation Model: a Framework for Evaluation of Interactive Information Retrieval Systems. *Information Research*, 8(3), paper no. 152. 2.1, 6.1.2

Brown, J. S. and Duguid, P. (2000). *The Social Life of Information*. Harvard Business School Press. 8

Buchanan, G., Blandford, A., Jones, M., and Thimbleby, H. (2004). Integrating information seeking and structuring: exploring the role of spatial hypertexts in a Digital Library. *Proc. ACM Hypertext & Hypermedia 2004*. 225–234. DOI: 10.1145/1012807.1012864 5.2.3

Burchard, J. E. (1965). How humanists use the library. In Overhage, C. F. J., and Harman, J. R. (Eds.) *Intrex: Report on a planning conference and information transfer experiments*. Cambridge, Mass: M.I.T. Press. 3.1.1

Campos, J. and Figueiredo, A. D. (2002). Programming for Serendipity. *AAAI Fall Symposium on Chance Discovery*. 48–60. 5.1.1

Carroll, J. M. and Rosson, M. B. (1992). Getting around the task-artifact cycle: how to make claims and design by scenario. *ACM Transactions on Information Systems*, 10(2), 181–21. DOI: 10.1145/146802.146834 2.4, 3.3, 6

Charmaz, C. (2006). *Constructing Grounded Theory: A practical guide through qualitative analysis*. Wiley. 4.1

Chi, E. H., Pirolli, P., Chen, K., and Pitkow, J. (2001). Using information scent to model user information needs and actions and the Web. In *Proc. SIGCHI Conference on Human Factors in Computing System*. ACM, New York, NY, 490–497. DOI: 10.1145/365024.365325 2.2

Chu, Y-C, Bainbridge, D., Jones, M. and Witten, I. (2004) Digital trail libraries. *Proc. JCDL*. ACM. 63-71. 2.4

Cooper, A. (1999) *The Inmates are Running the Asylum*. Indianapolis: Sams Publishing. 6.2.8

Cothey (2002). Longitudinal study of World Wide Web users' information searching behaviour. *Journal of the American Society for Information Science and Technology*. 53(2), 67–78. DOI: 10.1002/asi.10011 3.3.2, 4.6

Daniels, A. K. (1987). Invisible Work. *Social Problems*, 34(5). 403–415. DOI: 10.1525/sp.1987.34.5.03a00020 1

Davis-Perkins, V., Butterworth, R., Curzon, P., and Fields, B. (2005). A study into the Effect of Digitastion Projects on the Management and Stability of Historic Photograph Collections. *ECDL 2005*. 278–289. 3.1.2

Dervin, B., (1983). An Overview of Sense-making Research: Concepts, Methods and Results. Paper presented at the *Annual Meeting of the Int. Communication Assoc.* Dallas, TX. 2.3

Dervin, B. and Nilan, M. (1986). Information needs and uses. *Annual Review of Information Science and Technology*, 21, 3–33. 2.2

Diriye, A., Blandford, A., and Tombros, A. (2009). A polyrepresentational approach to interactive query expansion. In *Proc. 9th ACM/IEEE-CS Joint Conference on Digital Libraries*. ACM, New York, NY, 217–220. DOI: 10.1145/1555400.1555434 5.2.1

Druin, A. (2005). What children can teach us: Developing digital libraries for children. *Library Quarterly* , 75(1), 20–41. DOI: 10.1086/428691 6.1.3

Duguid, P. (1996). Material Matters: The past and futurology of the book. In Nunberg (Ed.) *The Future of the Book*. Berkley & Los Angeles. University of California Press. 63–102. 3.1.2, 7

Dumais, S., Cutrell, E., Cadiz, J., Jancke, G., Sarin, R., and Robbins, D. C. (2003). Stuff I've seen: a system for personal information retrieval and re-use. In *Proc. 26th Annual international ACM SIGIR Conference on Research and Development in informaion Retrieval*. ACM, New York, NY, 72–79. DOI: 10.1145/860435.860451 2.6

Duncker, E. (2002). Cross-cultural usability of the library metaphor. In *Proceedings of JCDL'02*, ACM Press. 223–230. DOI: 10.1145/544220.544269 3.1.1

Ellis, D. (1989). A behavioural approach to information retrieval system design. *Journal of Documentation*, 45(3), 171–212. DOI: 10.1108/eb026843 2.2

Ellis, D., Cox, D., and Hall, K. (1993). A Comparison of the Information Seeking Patterns of Researchers in the Physical and Social Sciences, *Journal of Documentation* 49(4), 356–369 DOI: 10.1108/eb026919 2.2, 2.5

Ellis, D. and Haugan, M. (1997). Modelling the Information Seeking Patterns of Engineers and Research Scientists in an Industrial Environment. *Journal of Documentation*, 53(4), 384–403. DOI: 10.1108/EUM0000000007204 2.2, 2.5, 4.1

Ellis, D. and Oldman, H. (2005). The English literature researcher in the age of the Internet. In *Journal of Information Science*, 31, 29–36. DOI: 10.1177/0165551505049256 2.5

Erdelez, S. (2004). Investigation of information encountering in the controlled research environment. *Information Processing and Management*. 40. 1013–1025. DOI: 10.1016/j.ipm.2004.02.002 4.4, 5.1, 5.1.1, 6.1.2

Faisal, S., Cairns, P., and Blandford, A. (2006). Developing User Requirements for Visualizations of Literature Knowledge Domains. In *Proc. IV06*. DOI: 10.1109/IV.2006.42 5.2.2, 6.2

Faisal, S., Cairns, P., and Blandford, A. (2007). Building for Users not for Experts: Designing a Visualization of the Literature Domain. In *Proc. IV07*. DOI: 10.1109/IV.2007.32 5.2.2

Farooq, U., Ganoe, C. H., Carroll, J. M., Councill, I. G., and Giles, C. L. (2008). Design and evaluation of awareness mechanisms in CiteSeer, *Information Processing & Management*, 44(2), 596–612. DOI: 10.1016/j.ipm.2007.05.009 5.1.2, 7

Fields, B., Keith, S., and Blandford, A. (2004). Designing for Expert Information Finding Strategies. In S. Fincher, P. Markopoulos, D. Moore, and R. Ruddle (Eds.) *People and Computers XVIII – Design for Life, Proc. HCI04*. Springer. 89–102. DOI: 10.1007/1-84628-062-1_6 3.2.1, 3.3.2

Flower, L. and Hayes, J.R. (1981). A cognitive process theory of writing. *College Composition and Communication*, 32, 365–387. DOI: 10.2307/356600 3.3.1

Foster, A. E. and Ford, N. (2003). Serendipity and information seeking: an empirical study. *Journal of Documentation*. 59. 321–340. DOI: 10.1108/00220410310472518 4.4, 7

Fox, N. J., Ward, K. J., and O'Rourke, A. J. (2005). The 'expert patient:' empowerment or medical dominance? The case of weight loss, pharmaceutical drugs and the Internet. *Social Science & Medicine*. 60. 1299–1309. 3.2.3

Glaser, B. and Strauss, A. (1967). *The discovery of grounded theory: Strategies for qualitative research*. Chicago, Aldine 6.1.5

Gristock, J. and Mansell, R. (1998). Distributed library futures: IT applications for 2000 and beyond. discussion paper at the *IATUL Conference, (The challenge to be relevant in the 21st Century)*, University of Pretoria, Pretoria, South Africa, (1–5). 3.2.1

Harrison, S. and Dourish, P. (1996). Re-Placeing Space: The Roles of Space and Place in Collaborative Systems. *Proc. ACM Conf. Computer-Supported Cooperative Work CSCW'96* (Boston, MA), 67–76. New York: ACM. DOI: 10.1145/240080.240193 3.1.1

Hartson, H.R. and Hix, D. (1993). *Developing User Interfaces*, John Wiley, New York. 6

Hartson, H. R., Shivakumar, P., and Púrez-Quiñones, M. A. (2004). Usability inspection of digital libraries: a case study. *International Journal of Digital Libraries*. 4(2). 108–123. DOI: 10.1007/s00799-003-0074-4 2.4, 3.2.2

Hearst, M. A. (2009). *Search User Interfaces*. Cambridge University Press. 8

Hsieh-Yee, I. (1993). Effects of Search Experience and Subject Knowledge on the Search Tactics of Novice and Experienced Searchers. *Journal of the American Society for Information Science*. 44(3). 161–174. DOI: 10.1002/(SICI)1097-4571(199304)44:3<161::AID-ASI5>3.0.CO;2-8 3.3.2, 6.1.2

Hyams, J. and Sellen, A. (2003). How knowledge workers gather information from the Web: Implications for peer-to-peer file sharing tools. *Proceedings of the British HCI Conference*, Bath, UK, Sept. 2003. 2.6

Ingwersen, P. and Järvelin, K. (2005). *The Turn: Integration of Information Seeking and Retrieval in Context (The Information Retrieval Series)*. Springer-Verlag New York, Inc. 2

Jones, W., Dumais, S. and Bruce, H. (2002). Once found, what then?: A study of 'keeping' behaviors in the personal use of web information. *Proceedings of ASIST 2002*, Philadelphia, Pennsylvania. DOI: 10.1002/meet.1450390143 2.6

Klein, G., Moon, B. and Hoffman, R. (2006). Making Sense of Sensemaking 2: A Macrocognitive Model. *IEEE Intelligent Systems*. 21(5), 88–92. DOI: 10.1109/MIS.2006.100 2.3

Klein, G., Phillips, J.K., Rall, E.L. and Peluso, D.A. (2007). A Data-frame Theory of Sensemaking. In Expertise Out of Context, *Proc. of the Sixth International Conf. on Naturalistic Decision Making*. Lawrence Erlbaum Assoc. Inc, US, 113–155. 2.3

Kubinyi, H. (1999). Chance favors the prepared mind – from serendipity to rational drug design. *Journal of Receptor and Signal Transductions Research*. 19. 15–39. DOI: 10.3109/10799899909036635 4.4

Kuhlthau, C.C. (1991). Inside the search process: information seeking from the user's perspective, *Journal of the American Society for Information Science*, 52(5), 361–371. DOI: 10.1002/(SICI)1097-4571(199106)42:5%3C361::AID-ASI6%3E3.0.CO;2-%23 2.2, 3.3.1, 4.6

Kuhlthau, C. C. and Tama, S. L. (2001). Information search process of lawyers, a call for 'just for me' information services. *Journal of Documentation*, 57(1), 25–43. DOI: 10.1108/EUM0000000007076 2.4, 2.5, 3.3.1, 6.1.1

Kuniavsky, M. (2003). *Observing the user experience: a practitioner's guide to user research*, Morgan Kaufmann, San Francisco. 6.1.4

Lankes, R.D. (2004) The Digital Reference Research Agenda. *JASIST*, 55(4), 301–311. DOI: 10.1002/asi.10374 3.2.1

Lawson B. (1997). *How designers think: The design process demystified*, 3rd ed. Oxford, Butterworth-Heinemann. 4.6

Lipson, J. G. (1997). The politics of publishing: protecting participants' confidentiality. In J. Morse (Ed.) *Completing a qualitative project: details and dialogue*. Sage Publications. 6.1.3

Livingstone, S., and Bober, M. (2004). Taking up opportunities? Children's uses of the internet for education, communication and participation. *E-Learning*, 1(3), 395–419. DOI: 10.2304/elea.2004.1.3.5 6.1.3

Mackay, W.E. (1995). Ethics, lies and videotape... . *Proceedings of the ACM conference on Human Factors in Computing Systems (CHI '95)*, ACM Press, 138–145. 6.1.3

Mahoui, M. and Cunningham, S. J. (2000). A comparative transaction log analysis of two computing collections. In *proceedings of ECDL'00*. Heidelberg: Springer. 418–423. DOI: 10.1007/3-540-45268-0_53 2.4

Makri, S., Blandford, A., and Cox, A.L. (2008). Investigating the information-seeking behaviour of academic lawyers: From Ellis's model to design. *Information Processing and Management* 44(2), 613–634. DOI: 10.1016/j.ipm.2007.05.001 2.5

Makri, S, Blandford, A., Gow, J., Rimmer, J., Warwick, C., and Buchanan, G. (2007). A Library or Just another Information Resource? A Case Study of Users' Mental Models of Traditional and Digital Libraries. *Journal of the American Society of Information Science and Technology*. 58(3). 433–445. DOI: 10.1002/asi.20510 3.2.1, 3.3.2

Marchionini, G. (1995). *Information seeking in electronic environments*. Cambridge University Press. 2.2

McAdam, R. and McCreedy, S. (2000). A critique of knowledge management: using a social constructionist model. *New technology, work and employment*. 15(2). 155–168. DOI: 10.1111/1468-005X.00071 2.6

McCown, F., Bollen, J., and Nelson, M.L. (2005). Evaluation of the NSDL and Google for Obtaining Pedagogical Resources. *Proc. ECDL 2005*. Lecture Notes in Computer Science, Volume 3652, 344–355. 2.4, 6.1.1

Miles, M. B. and Huberman, A. M. (1994). *Qualitative Data Analysis: An Expanded Sourcebook*. Sage Publishers. 6.1.5

Morris, S., Morris, A. and Barnard, K. (2004) Realistic books: a bizarre homage to an obsolete medium? *Proc. JCDL*. ACM. 78-86. 2.4

Nardi, B and O'Day, V. (1996). Intelligent Agents: What we learned at the library. *Libri*, 46, 59–88. DOI: 10.1515/libr.1996.46.2.59 3.2.1

Neuwirth C. and Kaufer D. (1989). The role of external representations in the writing process: Implications for the design of hypertext-based writing tools. *Proceedings of Hypertext '89*, 319–341. DOI: 10.1145/74224.74250 3.3.1

Nicholas, D., Huntington, P., Jamali, H. R. and Watkinson, A. (2006). The information seeking behaviour of the users of digital scholarly journals, *Information Processing & Management*, 42(5), 1345–1365. DOI: 10.1016/j.ipm.2006.02.001 6.1.4

Nielsen, J.(1994). Heuristic evaluation. In J. Nielsen and R. Mack (Eds.), *Usability Inspection Methods*, New York: John Wiley, 25–62. DOI: 10.1145/223355.223730 2.4, 6

O'Hara, K.P., Taylor, A., Newman, W., and Sellen, A.J. (2002). Understanding the materiality of writing from multiple sources, *International Journal of Human-Computer Studies*, 56, 269–305. DOI: 10.1006/ijhc.2001.0525 3.1.2, 3.3.1

Pagano, R. (2001). *Understanding Statistics in the Behavioral Sciences*, 6th edn, Wadsworth. 6.1.5

Palmer, C. L. (1999). Structures and strategies of interdisciplinary science. *Journal of the American Society for Information Science*, 50(3), 242–253. DOI: 10.1002/(SICI)1097-4571(1999)50:3<242::AID-ASI7>3.0.CO;2-7 2.5

Palmer, C. L., Teffeau, L. C., and Pirmann, C. M. (2009). *Scholarly Information Practices in the Online Environment: Themes from the Literature and Implications for Library Service Development*. Dublin, OH: OCLC Research and Programs. 2.5

Pirolli, P. and Card, S. (1995) Information foraging in information access environments. In I. R. Katz, R. Mack, L. Marks, M. B. Rosson, and J. Nielsen, Eds. *Proc. ACM CHI*. ACM Press/Addison-Wesley Publishing Co., New York, NY, 51–58. 2.2, 2.3, 4.5

Pirolli, P. and Card, S., (2005). The Sensemaking Process and Leverage Points for Analyst Technology as Identified Through Cognitive Task Analysis. In *Proc. International Conference on Intelligence Analysis*, McLean, VA, May 2–6, 2005. 2.3, 4.3, 5.2.2

Prabha, C., Connaway, L. S., Olszewski, L., and Jenkins, L. (2007). What is enough? Satisficing information needs. *Journal of Documentation*, 63, 74–89. DOI: 10.1108/00220410710723894 4.6

Rasmussen, J. Pejtersen, A.M., and Goodstein, L.P. (1994). *Cognitive Systems Engineering*, New York, Wiley. 4.1

Rimmer, J., Warwick, C., Blandford, A., Gow, J., and Buchanan, G. (2008). An examination of the physical and digital qualities of humanities research. *Information Processing and Management.* 44(3). 1374–1392. 3.1.1, 3.1.2, 3.2.1, 3.2.2

Rodden, K. and Fu, X. (2007) Exploring how mouse movements on Web Search results pages, *Proceedings of ACM SIGIR 2007 Workshop on Web Information Seeking and Interaction.* 29–32. 3.1.2

Robertson, S. (2008) On the history of evaluation in IR., *Journal of Information Science*, 34: 439-456. 2.1

Russell, D. M., Stefik, M. J., Pirolli, P. and Card, S. K. (1993). The Cost Structure of Sensemaking. In *Proc. of the INTERACT '93 and CHI '93 Conf. on Hum. Factors in Comp. Sys*, ACM Press, New York, NY, 269–276. DOI: 10.1145/169059.169209 2.3

Sackett, D. L. (1997). *Evidence-based Medicine: How to Practice and Teach EBM.* Churchill Livingstone. 3.3

Savolainen, R. (2006). Information Use as Gap-bridging: The Viewpoint of Sense-Making Methodology. *Journal of the American Society for Information Science and Technology*, 57(8), 1116–1125. DOI: 10.1002/asi.20400 2.3

Schön, D.A. (1983). *The Reflective Practitioner: How Professionals Think in Action*, Basic Books, New York. 4.6

Sellen, A. J., and Harper, R. H. R. (2002). *The myth of the paperless office.* MIT Press, Cambridge. 3.1.2

Sellen, A., Murphy, R., and Shaw, K. (2002). How Knowledge Workers Use the Web. *Proceedings of CHI 2002*, Minneapolis, MN. New York: ACM Press, 227–234. DOI: 10.1145/503376.503418 2.6

Sharp, H. Rogers, Y., and Preece, J. (2007). *Interaction Design* (2nd Edition). Wiley, New York 6, 6.1

Sharples, M. (1996). An account of writing as creative design. In Levy C.M. and Ransdell S. (eds.), *The science of writing.* New Jersey, Lawrence Erlbaum. 3.3.1, 4.6

Shen, R., Vemuri, N., Fan, W. da S. Torres, R. and Fox, E. (2006) Exploring digital libraries: integrating browsing, searching and visualisation. *Proc. JCDL.* ACM. 1-10. 2.4

Shneiderman, B. (2000). Creating creativity: user interfaces for supporting innovation. *ACM Transactions on Computer-Human Interaction* 7(1), 114–138. DOI: 10.1145/344949.345077 3.3.1

Sillince, E., Briggs, P., Fishwick, L., and Harris, P. (2004). Trust and mistrust of online health sites. *Proc. ACM CHI 2004.* 663–670. DOI: 10.1145/985692.985776 3.2.1, 4.2.1

Smith, P. J., S. J. Shute, D. Galdes and M. H. Chignell (1989). Knowledge-based search tactics for an intelligent intermediary system. *ACM Transactions on Information Systems* 7(3), 246–270. DOI: 10.1145/65943.65947 3.3.2

Spence, R. (1999). A Framework for Navigation. Int. J. of Hum.-Comp. St. 51, 919–945. DOI: 10.1006/ijhc.1999.0265 2.3

Stasko, J., Görg, C., and Liu, Z. (2008). Jigsaw: Supporting Investigative Analysis through Interactive Visualization, Information Visualization, 7(2), 118–132. DOI: 10.1057/palgrave.ivs.9500180 2.3, 5.2.2

Stelmaszewska, H. and Blandford, A. (2002). Patterns of interactions: user behaviour in response to search results. In A. Blandford and G. Buchanan (eds.) *Proc. Workshop on Usability of Digital Libraries at JCDL'02.* 29–32. 3.3.2, 6.1.1

Stelmaszewska, H. and Blandford, A. (2004). From physical to digital: a case study of computer scientists' behaviour in physical libraries. *International Journal of Digital Libraries.* 4(2). 82–92. DOI: 10.1007/s00799-003-0072-6 3.1.1, 3.3.2

Stelmaszewska, H., Blandford, A., and Buchanan, G. (2005). Designing to change users' information seeking behavior: a case study. In S. Chen and G. Magoulas (Eds.) *Adaptable and Adaptive Hypermedia Systems.* 1–18. London: Information Science Publishing. DOI: 10.1007/s10209-006-0053-9 4.1

Strauss, A, and Corbin J. (1998). *Basics of Qualitative Research: Techniques and Procedures for Developing Grounded Theory.* 2nd ed. Sage Publications, Inc. 4.1

Susskind, R. (2008). *The End of Lawyers? Rethinking the nature of legal services.* Oxford University Press. 3.2.3

Sutcliffe, A . and Ennis, M. (2000). Designing intelligent assistance for end-user information retrieval. *Proc. OZCHI00* 202–210. 3.3.2

Tague-Sutcliffe, J. (1992). The pragmatics of Information Retrieval Experimentation, Revisited. *Information Processing and Management.* 28(4), 467–490. DOI: 10.1016/0306-4573(92)90005-K 2.1

Taylor, R. S. (1968). Question-negotiation and Information Seeking in Libraries. *College & Research Libraries,* 29(3), 178–194. 2.2, 3.2.1

Theng, Y. L. (2002). Information Therapy in Digital Libraries. In *Proceedings of ICADL'02, Digital Libraries : People, Knowledge and Technology.* Heidelberg: Springer. 452–464. DOI: 10.1007/3-540-36227-4_53 3.2.1

Theng, Y. L., Mohd-Nasir, N., Buchanan, G., Fields, B., Thimbleby, H. and Cassidy, N. (2001). Dynamic Digital Libraries for Children. *First ACM and IEEE Joint Conference in Digital Libraries*, Roanoke (Virginia), 406–415. DOI: 10.1145/379437.379738 6.1.3

Thomas J.B., Clark, S.M. and Gioia, D.A. (1993). Strategic Sensemaking and Organisational Performance: Linkages among Scanning, Interpretation, Action and Outcomes. *Academy of Management Journal*, 36, 239–270. DOI: 10.2307/256522 2.3

Toms, E. (2000). Serendipitous Information Retrieval. *Proceedings of the First DELOS Network of Excellence Workshop in Information Seeking, Searching and Querying in Digital Libraries* (available from http://www.ercim.org/publication/ws-proceedings/DelNoe01/3_Toms.pdf). 4.4, 5.1.1

Toms, E. and McCay-Peet (2009). Chance Encounters in the Digital Library. *Proc. European Conference on Digital Libraries* (to appear). DOI: 10.1007/978-3-642-04346-8_20 5.1.1, 7

Twidale, M. B., Gruzd, A. A., Nichols, D. M. (2008). Writing in the library: Exploring tighter integration of digital library use with the writing process, *Information Processing & Management*, 44(2), 558–580. DOI: 10.1016/j.ipm.2007.05.010 3.1.1, 5.1.2, 5.3.1, 7

Vakkari, P. (2001). A theory of the task-based information retrieval process: a summary and generalisation of a longitudinal study. *Journal of Documentation*. 57(1), 46–60. DOI: 10.1108/EUM0000000007075 3.3.1, 3.3.2

Vakkari, P. (2003). Task-based information searching. *Annual Review of Information Science and Technology*, vol. 37 (B. Cronin, Ed.) Information Today: Medford, NJ, 413–464. 1

van Rijsbergen, C. J. (2001). Getting into information retrieval. In *Lectures on information Retrieval*, M. Agosti, F. Crestani, and G. Pasi, Eds. Springer Lecture Notes In Computer Science Series, vol. 1980. Springer-Verlag New York, New York, NY, 1–20. 2.1

Vishik, C. and Whinston, A. (1999). Knowledge sharing, quality and intermediation. In *proceedings of WACC*. ACM Press. 157–166, San Francisco. DOI: 10.1145/295665.295683 3.2.1

Warwick, C., Rimmer, J., Blandford, A., Gow, J., and Buchanan, G. (2009) Cognitive economy and satisficing in information seeking: A longitidunal study of undergraduate information behavior. *Journal of the American Society of Information Science and Technology* 60, 2402–2415. 3.3.2, 4.6

Weick, K. (1995). *Sensemaking in Organisations*. Sage, London, England. 2.3

Wenger, E. (1999). *Communities of Practice*. Cambridge University Press. 3.2.2, 4.1, 6.1.5

Wharton, C., Rieman, J., Lewis, C., and Polson, P. (1994). The cognitive walkthrough method: A practitioner's guide. In J. Nielsen and R. Mack (Eds.), *Usability inspection methods*, New York: John Wiley, 105–140. DOI: 10.1145/223355.223730 2.4, 6

White, R. W., Jose,, J. M., and Ruthven, I. (2006). An implicit feedback approach for interactive information retrieval. *Information Processing & Management*, 42(1), 166–190. DOI: 10.1016/j.ipm.2004.08.010 2.1

White, R. and Horvitz, E. (2009). Cyberchondria: studies of the escalation of medical concers in Web search. *ACM Trans. Inf. Syst.* 27, 4, 1–37. 3.2.3

White, R. and Roth, R. (2009). *Exploratory Search: Beyond the Query–Response Paradigm.* Synthesis Lectures on Information Concepts, Retrieval, and Services. Morgan & Claypool. 5.2.1, 8

Wilson, T. (2002). The nonsense of 'knowledge management.' *Information Research*, 8(1). 2.6

Wu, M. and Liu, Y. (2003). Intermediary's information seeking, inquiring minds, and elicitation styles. In *Journal of the American Society for Information Science and Technology*, 54(12), 1117–1133. DOI: 10.1002/asi.10323 3.2.1

Zhang, X., Qu Y., Lee Giles C. and Soong P (2008). CiteSense: Supporting sensemaking of research literature, *Chi 08*. DOI: 10.1002/asi.10323 2.3

Authors' Biographies

ANN BLANDFORD

Ann Blandford is Professor of Human–Computer Interaction in the Department of Computer Science at UCL, and Director of UCL Interaction Centre. Her research focuses on amplifying human capabilities through design, including work on minimising cognitive slips and on sensemaking. An important focus of her work has been on the use and usability of Digital Libraries, taking a multi-disciplinary approach that considers the design of technology, the knowledge and motivations of users, and the context within which information work takes place to develop a rich understanding of the design and use of such systems. She has been technical Chair for conferences including IHM-HCI 2001, HCI 2006, and NordiCHI 2010, and she has edited special issues of the International Journal of Digital Libraries and Information Processing and Management.

SIMON ATTFIELD

Simon Attfield is a Senior Research Associate and Lecturer at UCL Interaction Centre, University College London, where he has worked since receiving his PhD in 2005. His research interests lie in understanding how people think and work with information and what this means for the design of human-centred interactive systems. He has conducted numerous field and lab studies in information interaction, including studies of national news organizations (The Times, ITN), legal firms (Richards Butler, Freshfields Bruckhaus Deringer) and NHS healthcare settings. He has also consulted to news, legal and medical information providers; published internationally in numerous peer-reviewed academic outlets; presented research internationally to both academic and commercial audiences; and served on review committees for numerous international journals, conferences and workshops.

640941

Breinigsville, PA USA
09 December 2010
250945BV00005B/8/P